GREEN HOUSE GARDENING

How To Grow Amazingly Beautiful Greenhouse Plants To Die For

EVA CLAY

Copyright © 2020 Eva Clay

All Rights Reserved

Copyright 2020 By Eva Clay - All rights reserved.

The following book is produced below with the goal of providing information that is as accurate and reliable as possible. Regardless, purchasing this eBook can be seen as consent to the fact that both the publisher and the author of this book are in no way experts on the topics discussed within and that any recommendations or suggestions that are made herein are for entertainment purposes only. Professionals should be consulted as needed prior to undertaking any of the action endorsed herein.

This declaration is deemed fair and valid by both the American Bar Association and the Committee of Publishers Association and is legally binding throughout the United States.

Furthermore, the transmission, duplication or reproduction of any of the following work including specific information will be considered an illegal act irrespective of if it is done electronically or in print. This extends to creating a secondary or tertiary copy of the work or a recorded copy and is only allowed with express written consent from the Publisher. All additional right reserved.

The information in the following pages is broadly considered to be a truthful and accurate account of facts and as such any inattention, use or misuse of the information in question by the reader will render any resulting actions solely under their purview. There are no scenarios in which the publisher or the original author of this work can be in any fashion deemed liable for any hardship or damages that may befall them after undertaking information described herein.

Additionally, the information in the following pages is intended only for informational purposes and should thus be thought of as universal. As befitting its nature, it is presented without assurance regarding its prolonged validity or interim quality. Trademarks that are mentioned are done without written consent and can in no way be considered an endorsement from the trademark holder.

Table of Contents

PART I 8

Greenhouse Gardening 9

    Chapter 1: Siting and Building an Affordable Greenhouse 9

    Chapter 2: Plant Varieties for Greenhouses and Hothouses 13

    Chapter 3: Tips and Techniques for Easy Maintenance 20

    Chapter 4: Crop Rotation for Year-Round Growing 24

    Chapter 5: Structural Upkeep for Long-term Use 27

PART II 33

Traditional Vegetable Gardening 34

    Chapter 1: Designing and Building a Traditional Garden 34

    Chapter 2: Choosing Seeds and Plants for Beginners and Novices 40

    Chapter 3: Planting and Maintenance 52

        Sowing and Transplanting 53

        Watering 55

        Fertilizing and Composting 55

        Troubleshooting Weeds, Pests, and Disease 58

    Chapter 4: Harvesting and Preserving 62

    Chapter 5: Preparing for the Future 64

PART III 66

Raised Bed Gardening 67

    Chapter 1: Designing and Building a Raised Bed Garden 67

    Chapter 2: Choosing Plant Varieties That Thrive in Raised Beds 73

    Chapter 3: Easy Planting and Maintenance Techniques 83

    Planting Your Seedlings 85

    Watering for Maximum Plant Health 86

    Fertilizer and Compost for Optimum Growth 88

    Dealing with Weeds, Pests, and Pathogens 90

Chapter 4: Maximize Space with Vertical Growing 97

Chapter 5: Harvesting and Overwintering 98

PART IV 102

Urban Gardening 103

Chapter 1: Planning and Building an Urban Garden 104

Chapter 2: Choosing Plants for Food and Fun 109

Chapter 3: Planting and Maintenance Methods for Urban Settings 118

Chapter 4: Using Small Spaces to Your Advantage 125

Chapter 5: Planning for Continued Success 126

Chapter 6: Community Gardening Considerations 128

PART V 129

Chapter 1: What in The World Is A Thyroid 130

Chapter 2: Possible Thyroid Disorders 137

Hyperthyroidism 137

Hypothyroidism 144

Hashimoto's Disease 158

Graves' Disease 162

Goiter 167

PART VI 176

Renal Diet 176

Chapter 1: Easy Recipes for Managing Kidney Problems 177

Pumpkin Pancakes 178

Pasta Salad 180

Broccoli and Apple Salad 181

Pineapple Frangelico Sorbet 183

Egg Muffins 184

Linguine With Broccoli, Chickpeas, and Ricotta 186

Ground Beef Soup 189

Apple Oatmeal Crisp 190

Chapter 2: Weekend Recipes for Renal Diet 191

Hawaiian Chicken Salad Sandwich 191

Apple Puffs 192

Creamy Orzo and Vegetables 194

Minestrone Soup 196

Frosted Grapes 198

Yogurt and Fruit Salad 199

Beet and Apple Juice Blend 201

Baked Turkey Spring Rolls 202

Crab-Stuffed Celery Logs 204

Couscous Salad 205

Chapter 3: One-Week Meal Plan 207

Chapter 4: Avoiding Dialysis and Taking the Right Supplements 210

PART I

Greenhouse Gardening

Greenhouse gardening is a unique hobby that can produce stunning results for any level of gardener. Having a greenhouse means that you don't have to abide by some of the 'rules' of regular gardening because you can use a greenhouse to garden year-round and grow things that might not be considered common for your area. Although having a greenhouse can mean that you also have some additional maintenance chores, it's a rewarding undertaking for all those that attempt it.

Chapter 1: Siting and Building an Affordable Greenhouse

When you think about a greenhouse, you may think that they are too expensive to consider for your property, but there are a lot of amazing, affordable options available for would-be greenhouse gardeners today. You can build your own from a variety of materials, or you can get a pre-fabricated kit. You can also contract someone to do the construction for you if you're not comfortable with the building process. The first thing you need to do, though, is to find your site and decide on dimensions for your building. You should also check with your municipality to see if a greenhouse is considered a permanent or non-permanent structure, to determine if you need a variance or permit for the building.

Once you've determined that you can build a greenhouse and if there are any local restrictions on where you can place it, it's time to find your spot.

You want to set your greenhouse somewhere level, with plenty of sunlight, and near access to water and electricity. When you've found the perfect or largely perfect spot, measure it out and see how big a structure you can build. Keep in mind that you want something big enough to move around in, and small enough not to cost you a fortune to build. Most home greenhouses start at about 6'x9', and don't generally get too much larger than 14'x20'. A fairly standard dimension is 10'x14'.

Some materials to consider for your greenhouse structure are hardwood lumber and PVC vinyl piping if you are going to build your own greenhouse. You can use tempered glass, plexiglass, or heavy-duty plastic sheeting for the windows, and be sure to account for ventilation through vents or windows that can be opened. Building your own greenhouse can be rewarding if this is your forte, and plans are available for free or low-cost on the internet and in gardening magazines. More popular, however, are pre-fabricated kits, which are made to fit any budget. They come in all sizes and materials, so you can shop around and find what's best for you.

Sun's short waves

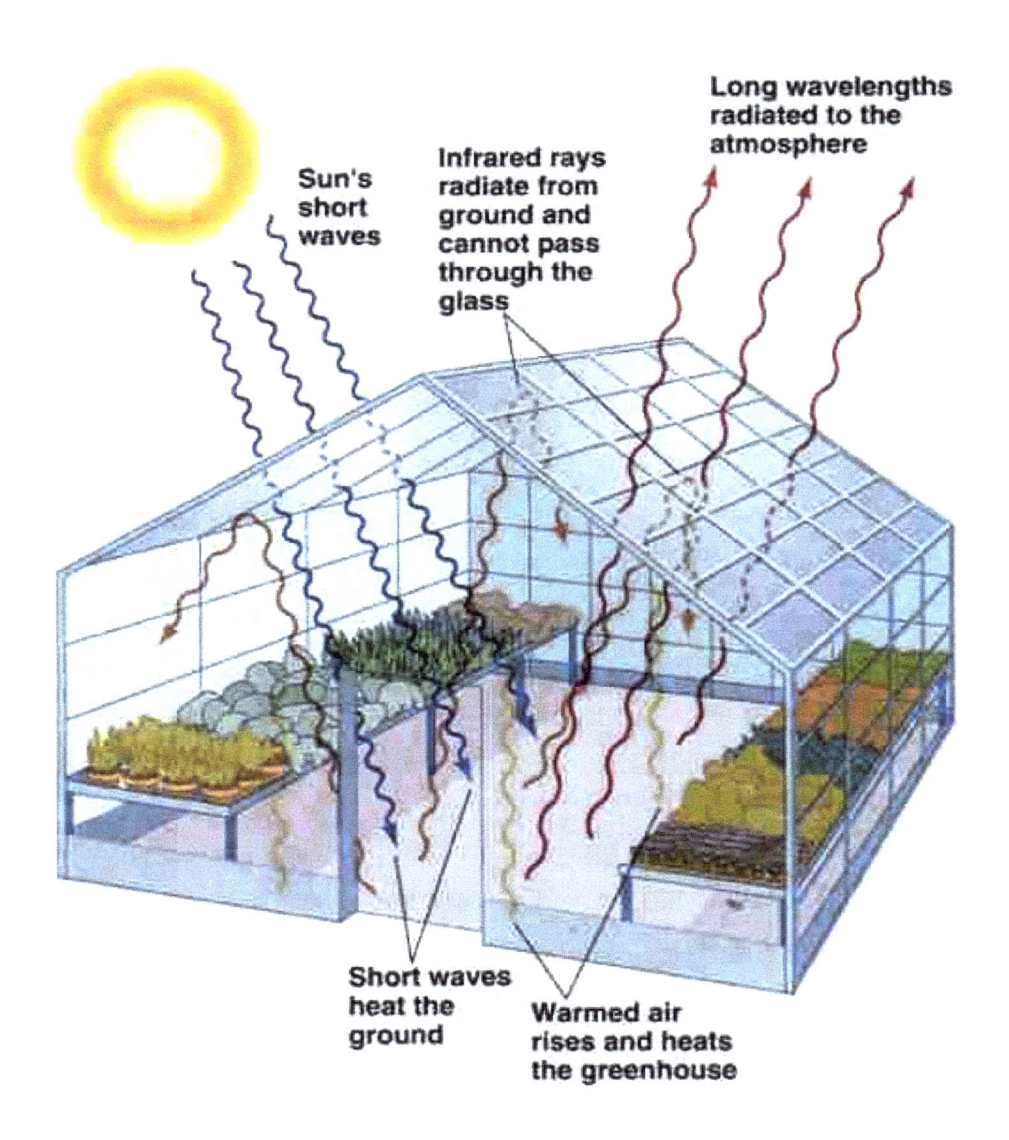

Infrared rays radiate from ground and cannot pass through the glass

Long wavelengths radiated to the atmosphere

Short waves heat the ground

Warmed air rises and heats the greenhouse

- Even without heating systems, greenhouses maintain warmer growing conditions -

Another thing to factor into the cost of erecting a greenhouse is the flooring materials. Are you going to level and pour a concrete slab, or do you think that gravel or QP (quarry process) stone is the better alternative?

This is entirely a personal decision, but when you check with your town's building or zoning officer, ask of concrete v. gravel makes a difference in whether a greenhouse is permanent or non-permanent. That might help you make your decision. Consider, too, if you'll need to spend time, money, and energy to level your property to accommodate the building.

The last expense to include when pricing out the cost of a greenhouse is water and electricity. If you need to run a supply from your house or garage to your greenhouse site, you should shop around with local electricians and plumbers to get a good value for your project. This isn't something you should attempt yourself unless you have extensive experience. If it's not too far, you can probably just run a hose and sturdy, exterior-use cord, but be aware that you'll probably have to wrap them up and bring them in every night. A dedicated underground feed is likely a better choice.

To outfit the inside of a greenhouse, let's talk briefly about the difference between a hothouse and a greenhouse. A hothouse is climate-controlled, meaning you have a heating and cooling system in place to maintain a desired temperature year-round. A greenhouse is not climate-controlled, meaning while it will extend your growing season in cooler climates, it may not still be a 12-month venture. If you're not up to investing in a full heating and cooling system, a floor model air conditioner and a space heater may be right up your alley. Consider adding an inexpensive humidifier if you live in a dry climate. Those three small appliances can usually be found cheaply or secondhand.

For greenhouse gardening, you're going to need a collection of containers, and how you choose them is entirely up to you. Some people like eclectic mismatched pots, and other like everything to match, only varying sizes to meet the needs of their plants. You'll also want some good hand tools to outfit your garden- a trowel, bulb digger, a hand rake, a pair of pruners, some utility scissors, and twine should get you set up nicely. You'll also need something to deliver your water to your plants, a sturdy watering can or a hose depending on the size of your greenhouse and your flooring. If you went with gravel, excess water would drain itself, but if you laid a concrete pad, you may want to get a long-handled squeegee to remove water from the floor and out the door.

The only other things you need for greenhouse gardening are a thermometer to keep tabs on your indoor temperature, buckets or baskets for waste and harvesting, some tool storage, and a garden journal to record your activities such as plant acquisition, major pruning, fertilizer applications, and the like. Having notes on your gardening practices helps you know what's working and what isn't, so you can make decisions for the future. If you're not into bending and twisting, you can put some tables to set your plants on and a stool to plant yourself on in the greenhouse. When you're all fitted out with tools, the last thing you'll need is a good supply of potting soil, and it's time to be a gardener!

Chapter 2: Plant Varieties for Greenhouses and Hothouses

You can grow almost anything in a greenhouse, and it's fun to decide what you'd like to cultivate. The beauty of a greenhouse, and especially a hothouse, is being able to grow things you might not be able to grow in your climate. Turning your structure into a green oasis makes it more than a place to work, it makes it a place to relax and unwind. Let's take a look at some of the plant varieties you can consider for their foliage, flowers, and food.

Philodendron- With nearly 500 varieties, you're sure to find something to love about these fast-growing vining plants. Philodendron has distinctive shiny foliage, and are easy to grow from nursery shoots in small- to medium-sized pots and hanging baskets. Philodendron will flower under warm, humid conditions, but are often kept for their foliage alone.

Alocasia- This is another plant in the same family as philodendron, and they come in a wide array of greens and patterns. Alocasia is considered to be tropical and subtropical, meaning they need warmth to thrive. They do well in greenhouses in medium pots but do require occasional re-potting as they are heavy feeders who will strip the nutrients from their soil. Purchase small seedlings from a nursery for the best results.

Geraniums- These flowering favorites do well in greenhouses year-round, and require a lot of sun and water. It's easy to buy a small flat of nursery shoots and transplant them into medium-sized pots that will provide interesting foliage all year and flowers in season. They are fragrant and come in a variety of colors.

Chrysanthemums- Outside of a greenhouse, mums are a fall staple in temperate climates. Inside a greenhouse, especially a hothouse, they can thrive all year. They come in nearly countless colors, and you can try starting them from seed or get seedlings. The delicate petals are beautiful to look at, and their faint aroma is earthy and pleasant. Mums tend to have shallow roots, so be sure to use deep pots and give them plenty of soil to avoid them 'heaving' themselves upward.

Orchids- Orchids are a classic hothouse flower, and they are popular all over the world. With their curved stalks and aromatic flowers, the cultivation of orchids can be a hobby of itself. You can try moth orchids, dendrobiums, and lady slippers to get your feet wet with this amazing species. You can encourage growth with orchid fertilizer and keep them happy by not over-watering them.

Growing Orchids

Orchids are among the most spectacular of all flowering plants. They can be planted in any container that provides adequate drainage and sufficient root ventilation.

Lighting

Most orchids require plenty of light, preferably at least six hours a day. Conversely, inadequate light prevents orchids from flowering, although they will grow.

Watering

Water orchids thoroughly, usually about once a week, and then allow them to dry slightly before watering again.

Humidity

Orchids require adequate humidity. Usually, around 60 percent or more is necessary. Use a humidifier, or set your orchids in a gravel-filled tray of water.

Roots System

Orchid roots are highly specialized organs designed to breathe and soak up water very quickly. They grow best with turbulent air circulation over their roots. The right potting mix for orchids provides plenty of drainage, air circulation or moisture depending on the needs of your particular orchids.

- Orchids are a popular plant that can become a hobby unto themselves -

Calathea- This is another great foliage plant and its varieties are often called 'prayer plants' due to the shape of the leaves, which come in a lot of different colors and patterns. These plants prefer a warmer environment but aren't too thirsty or hungry. Good soil and regular watering will keep them happy and growing. Move to larger pots to encourage larger growth, or keep them in medium pots to keep them in check, just mind that they don't get root-bound. Occasional transplanting will avoid this problem.

Nightshades- If you want to grow food in your greenhouse, it doesn't get much easier than the nightshade family, which includes tomatoes, peppers, eggplant, and potatoes. Yes! You can grow potatoes in a greenhouse if you've got a large enough container. Nightshades grow well in greenhouses because the environment keeps them from many of the insect pests that normally plague these species, and they love the warm air and sunshine. If you keep a hothouse, you can plant nightshades for food year-round and be able to enjoy fresh, juicy tomatoes in the dead of winter. Fun!

Be sure to use wide, deep pots and stake your plants so they stay upright. For best results in a greenhouse, get your nightshades from seedlings, and plant them in the container they will stay in for their whole lifecycle. You can also get seed potatoes at almost any nursery or online. Nightshades are heavy feeders, so start them out with good potting soil and give them a bump of organic plant food halfway through their growing cycle of when you see the first flowers start to appear. Don't overfeed, though. Be sure to follow package directions.

Legumes- Leguminous plants are fun to grow in a greenhouse for food, like beans and peas, and if you've got some deep containers, peanuts. The tiny, fragrant flowers on legumes are beautiful to see, as well. You can grow vining varieties of beans with stakes, or in your greenhouse corners, and they grow so quickly, you can have fresh crops year-round to eat and preserve through freezing, canning, and drying. If you're a fan of Mediterranean or Middle Eastern cuisine, you can try your hand at growing lentils and chickpeas to make your own falafel, hummus, and other dishes. Legumes grow well from either seed or seedling.

Cucurbits- This is the family of plants that includes cucumbers, melons, and squashes, and if you like them grown in a traditional garden, you'll love to grow them indoors. You can easily start cucurbits from seed or seedling, and they grow quickly as long as they have a hearty water supply. You can get compact container varieties of all your favorite cucurbits, or trellis and stake vining varieties to keep them upright and healthy. It's a really neat surprise to have a fresh watermelon on your winter holiday table!

Herbs- The hardiness and versatility of herbs make them perfect for growing in greenhouse containers. You can grow almost any herb imaginable in pots, and they provide aromatics, repel insects, and best of all, can go straight from the greenhouse to your kitchen with just a few snips. Try flowering varieties like bergamot, or discover the many cultivars of basil and oregano. The great thing about growing herbs indoors, in pots, is that the perennial species won't have the opportunity to start taking over your garden, as can sometimes happen in a traditional native soil plot. Play around with growing herbs from seeds and from nursery starts to see which you like better.

Ferns- If your greenhouse has some shade in the corners, you might want to consider growing some ferns. Ferns are a lot of fun because they come in such amazing shapes and colors. Ferns like to be warm and have nice wet soil, and they thrive in humid greenhouse environments. They don't, however, like full sun all day, so if you put them in the darkest corner of the building, they will be thrilled and reward you with terrific growth. If you want to grow some ferns, but don't have a shady corner, you can just cover the plants with a light-colored fabric during the hottest part of the day. Towels or light canvas works very well for this purpose.

These are just some of the wonderful 'starter' plants you can cultivate in a greenhouse environment. If you're considering other varieties, be sure to read the plant tags carefully before you purchase anything. Make sure that the seedlings are marked as 'compact' or 'good for containers', and take a close look at the watering and feeding requirements, as well as the spacing needs. You want to make sure you've got the right pots for your new seedlings to move into. As tempting as it can be, don't buy anything you can't provide the correct environment for.

Chapter 3: Tips and Techniques for Easy Maintenance

With your plants chosen and your greenhouse set up, you can get planting and arranging to your heart's content. The first thing you should do is set up your pots with soil, not filling them quite all the way. You want to have room to work without knocking soil all over the place. For pots where you are only placing one seedling or plant, you should make your planting space dead center, making a hole larger than the pot that the seedling came in.

You should be gentle when transplanting, giving the pot a good squeeze and, grasping the plant at the base of the stem, wiggling until the seedling comes loose. Holding it over its new container, give the roots a soft massage and placing the plant into the hole you've made. Then cover the roots and settle the soil in around the stem, pressing it lightly to hold the plant upright. For planting multiple seedlings in large containers, be sure to space them out enough so their roots have a chance to grow both down and out for stability. When your seedlings are planted, give them a good drink and let them settle into their new homes. Label your plants with their original tags or make them new ones with pertinent info.

Arrange your plants how you please in your greenhouse, but make sure not to place your containers too close to each other, as they will need airflow to stay healthy. Leave room for yourself to work and move around safely and comfortably. If you've hung work lights or have other electrical cords, make sure to tuck them away or run the cords under your tables so you don't have any tripping hazards. Safety is the name of the game when you have the potential to make a soil mess, or worse, hurt yourself!

To maintain your greenhouse plants, you should set a regular watering schedule and stick to it! Your plants will be happiest when they are well-watered. If you have any plants that need more or less frequent watering, be sure to adjust their schedule accordingly. Always water the roots of the plants, not the leaves, because wet leaves can invite infection. While weeds, pests, and pathogens are of lower concern in a greenhouse than outdoors, let's go over some basics for dealing with any issues so your plants can thrive and grow.

You can usually manually weed your pots for any intruders in a greenhouse, so just be vigilant about anything odd sprouting up in your pots. Weed seeds don't usually hide away in potting mix, but they can blow in when you've got the doors and windows open for ventilation. Take care of weeds promptly, and you won't have any undue issues. Being vigilant also helps you guard against any pests and pathogens, but the chance of invasion or infestation is never zero.

Insect pests have a way of creeping in no matter how secure your greenhouse is. Some common pests to look for are aphids and whiteflies, especially if you are growing vegetables. Take affected plants outdoors and give them a blast from the hose. The water pressure is enough to remove and drown the pests. Thrips, leaf miners, and snails and slugs also favor foliage plants, and you can treat your plants with horticultural soaps to get rid of the first two. For snails and slugs, get them drunk! Place low-sided dishes of beer near your plants at night, and in the morning, you'll have dead slugs and snails. They can't resist the lure of a good brew. Or a bad one; by all means, use up the old stuff someone left in your cooler after a BBQ last summer. Try to avoid the use of harsh insecticides in the enclosed space of your greenhouse.

Pathogens are a different issue, and you should learn to tell whether your plant has an infection or a nutrient imbalance. Your potting soil should be well-stocked with the important macronutrients (nitrogen, phosphorus, and potassium) when you first get your plants installed. As the roots take up those nutrients, heavy feeding plants can strip the nutrients from the soil. If you see signs of yellowing and curling, browning and wilting, your plant may need a food boost. You can use nutrient sticks or add compost to your containers to up the organic content.

If you see white or rusty spots on your foliage, you are likely dealing with a pathogen, not a nutrient issue. Other signs of disease include rotting roots or blossoms, leaf veins yellowing, and leaf drop. Common pathogen issues for greenhouse plants include powdery and downy mildew, leaf spot, anthracnose, and gray mold fungus. If you see signs of disease, prune away the affected tissue and dispose of it away from the greenhouse, and do not compost it. Always clean your tools with a disinfectant after cutting away diseased tissue. If you need help identifying any pests or pathogens, you should reach out to your local Cooperative Extension or Farm Bureau. They have trained staff and volunteers who can give you guidance on handling pests and disease.

Like other living beings, plants are more susceptible to issues when they are vulnerable. Keep their immune systems strong by using best gardening practices. This means good watering techniques, as well as being observant and proactive. Greenhouse plants, especially foliage plants, need a lot of TLC. You should refresh their soil with compost, or the occasional organic fertilizer, when needed. Good soil is the key to any garden but is so vital to plants in containers. It is their main source of nutrients, and you need to keep it viable to maintain plant health. Replace the soil every few years if you find that you cannot revitalize it with compost or fertilizer anymore.

You will also want to prune your foliage plants regularly to help them keep a desired shape or size, making sure to always prune at the joints to avoid 'open wounds' on the stems. You should also transplant your plants when they begin to outgrow their homes. You don't want them to become rootbound and be unable to take up nutrients. Having room to spread their roots can mean the difference between a happy plant and a distressed plant.

By being a patient, vigilant gardener, you can have healthy, thriving greenhouse plants that will reward you with beautiful foliage and flowers, as well as fresh food even in the cooler months. Using best gardening practices, asking for help when you need it, and tending to your plants' needs with gentle, knowledgeable action will give you a garden you can be proud of and want to work in every day.

Chapter 4: Crop Rotation for Year-Round Growing

If you've equipped your greenhouse to be a hothouse, you can use your space to grow year-round, which is fantastic! That means you'll be able to, if you like, do some crop rotation or succession planting, so that you don't always have the same things growing at the same time. You'll likely want to keep the foliage plants you've put so much effort into, but if you're growing vegetables, swap them out sometimes so you don't get bored with what you have.

Suitable for

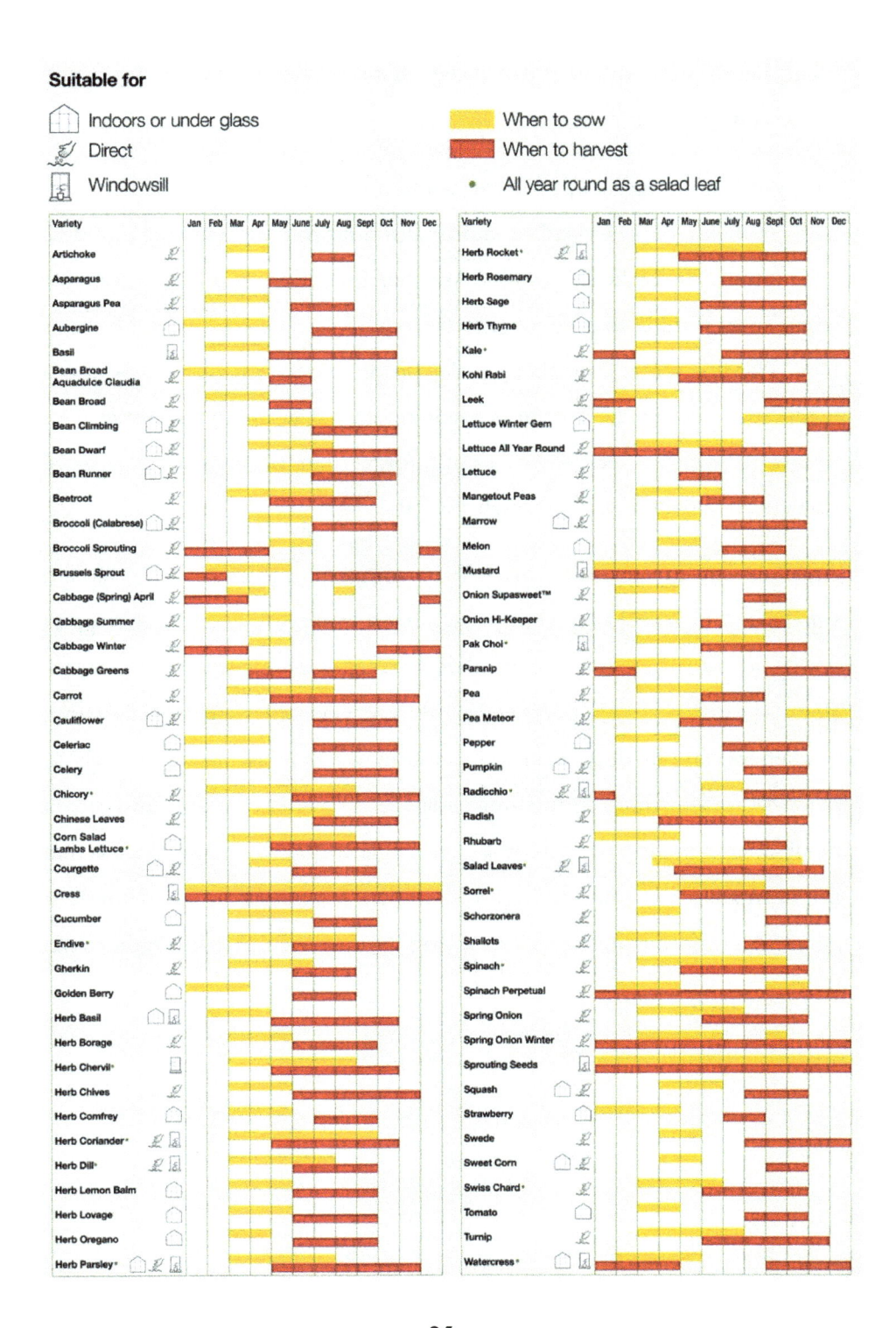

- 🏠 Indoors or under glass
- ✍ Direct
- 🪟 Windowsill
- 🟨 When to sow
- 🟧 When to harvest
- • All year round as a salad leaf

Variety	Jan	Feb	Mar	Apr	May	June	July	Aug	Sept	Oct	Nov	Dec
Artichoke												
Asparagus												
Asparagus Pea												
Aubergine												
Basil												
Bean Broad Aquadulce Claudia												
Bean Broad												
Bean Climbing												
Bean Dwarf												
Bean Runner												
Beetroot												
Broccoli (Calabrese)												
Broccoli Sprouting												
Brussels Sprout												
Cabbage (Spring) April												
Cabbage Summer												
Cabbage Winter												
Cabbage Greens												
Carrot												
Cauliflower												
Celeriac												
Celery												
Chicory*												
Chinese Leaves												
Corn Salad Lambs Lettuce*												
Courgette												
Cress												
Cucumber												
Endive*												
Gherkin												
Golden Berry												
Herb Basil												
Herb Borage												
Herb Chervil*												
Herb Chives												
Herb Comfrey												
Herb Coriander*												
Herb Dill*												
Herb Lemon Balm												
Herb Lovage												
Herb Oregano												
Herb Parsley*												

Variety	Jan	Feb	Mar	Apr	May	June	July	Aug	Sept	Oct	Nov	Dec
Herb Rocket*												
Herb Rosemary												
Herb Sage												
Herb Thyme												
Kale*												
Kohl Rabi												
Leek												
Lettuce Winter Gem												
Lettuce All Year Round												
Lettuce												
Mangetout Peas												
Marrow												
Melon												
Mustard												
Onion Supasweet™												
Onion Hi-Keeper												
Pak Choi*												
Parsnip												
Pea												
Pea Meteor												
Pepper												
Pumpkin												
Radicchio*												
Radish												
Rhubarb												
Salad Leaves*												
Sorrel*												
Schorzonera												
Shallots												
Spinach*												
Spinach Perpetual												
Spring Onion												
Spring Onion Winter												
Sprouting Seeds												
Squash												
Strawberry												
Swede												
Sweet Corn												
Swiss Chard*												
Tomato												
Turnip												
Watercress*												

- This handy guide will help you know what's in season for companion and succession planting -

You can keep in season with what would normally be growing at the time, or have melons in the dead of winter and butternut squash in the middle of summer. It's really up to you. Another fun thing to do is to cultivate some spring bulbs to give as gifts for occasions like Easter or Mother's Day. Planting some annuals to make flower arrangements with is also a lovely idea that will keep your gardening juices flowing and brighten your home or someone else's day. Don't forget that herb mixes and dried flowers also make terrific, thoughtful gifts!

Another way to maximize space is to companion plant your herbs and vegetables. That will keep things growing in every pot, even when veggies are done with their run and need to be replaced or succession planted. Because most herbs are perennials and most vegetables are annuals, those containers can pull double duty year-round. It can be tempting, though, to plant too many things in too small a space. Try to keep plants away from being directly on top of each other in a single container. A good rule is that if it seems like it's going to be too cramped, it probably is. Go with your gut, and always err on the side of giving more space per plant, rather than less.

Chapter 5: Structural Upkeep for Long-term Use

Greenhouse gardening is a big commitment, because not only do you have to take care of the plants inside, you also need to take care of the structure itself. Take care to check your building for any structural damage, especially after storms. Replace any cracked glass or plexiglass as soon as you can, and if you've got a plastic greenhouse, look for any tears in the sheeting. By taking care of small issues before they become big ones, you'll save yourself lots of time, money, and frustration.

If your greenhouse is built with a wooden frame, be sure to keep it painted or water-sealed for extra protection from the elements. For structures made of PVC piping, it can be a good idea to use caulk or another sealant at the joints to make sure that moisture stays out of the materials. Keep an eye on the floors and doors of your building, too. If you've got a concrete slab, seal up any cracks to prevent further damage. Gravel floors should be refreshed every so often when you can afford to do so. Check that your door(s) close snugly and fix any loose hinges.

For gardeners that live in areas that get snow or ice, it's important to have a system in place to keep that winter weather from damaging your greenhouse. Get a good long-handled roof snow brush, and promptly remove as much snow as you can, especially if it's wet heavy snow that can be common in most temperate climes. A little preventative maintenance goes a long way! Make sure you occasionally get on a ladder and check out the ceiling/roof of your greenhouse, too. You don't want to be surprised by any falling panels or leaky tears in the roof! It's all about maintaining safety.

Humidity Too Low	Humidity Too High
Wilting	Soft growth
Stunted plants	Increased foliar disease
Smaller leaf size	Nutrient deficiencies
Dry tip burn	Increased root disease
Leaf curl	Oedema
Increased infestation of spider mites	Edge burn (guttation)

- Don't forget to mind your humidity levels and keep your greenhouse well-vented all year long!-

By keeping your plants inside happy and healthy and your structure sound and sturdy, you are well on your way to being an experienced greenhouse gardener. Keep the temperature steady, keep your plants watered and fed, and keep the building in good condition, and you will have a garden and greenhouse to be proud of for many, many years. Have fun!

One of the greatest things, or perhaps, THE greatest thing about being a gardener is that it allows you a lot of freedom to explore and express your interests. Hobby gardeners wear a lot of hats- they are manual laborers, researchers, scientists, chefs, florists, readers, writers, and naturalists. Along with those roles comes the inevitable. You cannot help yourself but learn, even if you learn by failing. The wonderful thing about plants is that you can take a seed, nurture it, feed and water it, and watch it live its entire life cycle under your care, and then! Just when you are mourning the loss of a particularly abundant tomato plant or the dying off of your annual flowers, you remember, you get to do this again next year. You get to be a part of the renewal of spring. It's an amazing feeling. Just the sheer act of putting your hands in the soil can be calming; it awakens your senses and lowers your stress levels.

Use your off-seasons wisely. Read up on the plant species and varieties that you want to discover more about. Everyone who gardens eventually finds their niche. Some are obsessed with vegetables, some with soil health, and some devote their gardening efforts to growing stunning floral displays. You'll find your niche, too. Take the time to explore the many fascinating facets of gardening. It's fun to learn about new plants and catch up on the latest news in botany and the life sciences.

- Put your journal to good use in the off-season with notes and new designs
-

Another awesome byproduct of gardening is how much closer you will feel to your environment. When you spend time outdoors or in a greenhouse with clear walls, you can see and hear what's happening around you. There is birdsong and the buzzing of insects, and on hot days, you can almost hear your plants speaking to you as they stand tall in the sun and wave in the summer breeze. It's magical. Let it both transport you and ground you, and you'll wonder how you ever lived without your garden.

Gardening also teaches you to be creative in other parts of your life. You can learn to cook with new ingredients, or take up canning and food preservation. You can practice making flower arrangements, or drying flowers for crafts. Gardening begets other hobbies, and it's easy to branch out into beautiful artwork and crafts to share with your family and friends.

You can also leave your plants in the garden and capture them in other ways, like photography or painting. Let your garden be your muse in all things.

The point is, learning how to grow a basic garden is only the beginning for you. You can use the planning and execution skills learned in gardening to expand your horizons into so many other things. So be curious. Ask why certain pests attack certain plants. Figure out how to stop it from happening. Watch what birds come to your garden at particular times of the year, and wonder if they are beneficial. Be a voracious researcher. You'll soon discover that the more you know, the more you will want to know. It's empowering in a way few hobbies can be.

- Flex your gardening muscles in the winter with indoor herbs! -

Gardening can also be frustrating, but take heart. For every step back, there will be more steps forward. Sometimes, we can do everything correctly and still fail! Weather can be a harsh mistress, and invasive insects can cause harm before anyone even knows they exist. Be stronger than the setbacks and don't get discouraged. Even the best plant scientists in the world lose crops sometimes. And sometimes, they are just winging things, too. So have fun. Read. Take notes. Sit in your garden and watch the bumblebees do their pollination thing. Most of all, take time to smell the flowers. The gardening world is your oyster, crack it open and find all the pearls!

PART II

Traditional Vegetable Gardening

One of the most fulfilling things that any gardener can do is grow their own food. Cultivating a traditional garden, or kitchen garden, is a great way to put fresh healthy fruits and veggies on your table and in your pantry. By planting and caring for a kitchen garden, you can develop a rich understanding and appreciation of where your food comes from. A kitchen garden is also a great way to involve your children and grandchildren in food production and being a part of the food web, as well as teach the science of gardening and the ecosystem.

Chapter 1: Designing and Building a Traditional Garden

Traditional gardens are built in native soil, and so the first thing you will need to do is choose a site. The primary consideration for where to place your garden should be sunlight. Vegetables require full sun, and at the very minimum, they should receive 6 hours of sunlight a day. Another consideration for choosing the site of your garden include proximity to a water source because no one wants to lug heavy watering cans around every day, and the closer and more convenient your water source is, the more likely you are to water regularly. The last major site consideration is whether or not the site is level or can be leveled, and if run-off or erosion will be an issue.

When you're building a garden in native soil, you should begin by observing the area you've chosen for a while before breaking ground. Watch the site when it rains to see what the run-off pattern is, and look for any wildlife activity so you know what you'll be dealing with protecting your plants from. You should also get a soil test done to determine if you'll need to do any amendments before you plant. Your local Cooperative Extension or Farm Bureau will be able to help you out getting a low-cost test kit. These typically take about one to two business weeks for return results, giving you plenty of time to plan your garden while you wait.

Typically, traditional gardens are designed and built with squares and rectangles, but you don't have to stick to the norm. Designing your garden can be really fun, and all you need is some paper and pencils, a good eraser, and a straight-edge and/or compass. Measure the site you're considering, and draw it out to scale on a large piece of paper. Yes, you can also do this with one of several software applications, but there's something especially satisfying about drawing it on paper. It gives you a hands-on connection to your garden before you've even moved a shovelful of soil.

When you are choosing how to design the beds in your garden, think about what you realistically need. How much food do you want to produce? The general rule is that you need 200 sq. feet of garden space per person, assuming everyone in your family is good about eating their veggies. If you want to grow things to preserve, like cucumbers for pickles, you should consider that as well. Also, think about labor. Who will be caring for the garden? If it's only you, then plan accordingly for how much space you can maintain over the course of a growing season. There is no shame in starting small and adding to your garden in subsequent years. It's better to do that than start big, feel overwhelmed, fall behind, and either waste food, or worse, feel bad about yourself as a gardener.

You can create one big rectangle, but that's boring and can be difficult to maintain. Instead, consider squares- they are easy to work around and reach into, can be arranged in neat patterns, and can still be enclosed with one large fence to protect from wildlife. You can also consider circles, for the same reason. Be creative! You want to build a space you can enjoy for its beauty and functionality, not just its productivity. Think about other elements you can add to the design to make it a place you'll want to work and hang out. Maybe a birdbath, or a bench? How about a bistro table, so you can enjoy your morning coffee while you go about your weeding and watering routine? The possibilities are endless!

Sketch a few designs, and pin them up somewhere where you can see them regularly. I guarantee that within a few days of seeing those drawings on the refrigerator, one will become your favorite. The design you make with your head and love with your heart is always a good choice because you feel protective and nurturing of having created it. One of the greatest rewards of being a gardener is feeling proprietary of your space, knowing it's something you built with your own creativity and labor.

Once you've settled on the design you'll be using for your garden, it's time to break ground. For this step, you'll need a measuring tape, some stakes, twine, and a hammer, and whatever tool you've decided to use to break your soil. This could be hand tools, like a shovel, garden ax, and pitchfork, or a mechanical implement like a roto-tiller or small garden tractor. Your first step will be to measure and mark out your dimensions with your twine and stakes. Then double check-it so you know you're digging up the right spots! You want to till down and turn over the soil at least six to eight inches. When you are done, leave it alone for a few days to dry out and 'rest'. You should relax a bit too! Tilling is hard work, but it's fun to see your dream garden starting to take shape right before your eyes.

If you've gotten your soil test results back, your report will tell you everything you need to know about the nutrient content and pH of your soil, and if anything's not quite right, you'll be given specific instructions on how to amend it. You may find recommendations for adding lime, potash, fertilizer, or other amendments. If so, do this according to the 'dosages' on your report. If you are unsure how to do this or want to be certain you are reading the report correctly, you should contact the Farm Bureau or Cooperative Extension where you acquired your test kit and ask them to assist you. Soil amendments, even if you decide to use organic solutions, are nothing to mess around with. There can be too much of a good thing, and you don't want to cause toxicity in your garden soil.

Even if you don't have any recommended soil amendments, it's a good idea to prep your soil before planting with some compost to add organic matter. Healthy soil is a living, breathing entity, packed with microorganisms and insect life. When you add organic matter to your garden, you are providing nutrition to the underground ecosystem that will greatly benefit your plants. These organisms excrete nitrogen that plants need to thrive and enrich your soil structure by increasing pore space. This allows more room for roots to expand and lets the soil retain more water. You can never- let me emphasize- *never* have too much organic matter in your garden soil.

If you can get compost in bulk, that's awesome, but if you can't, bagged compost works well, too. You should also invest in a bag of good organic fertilizer granules. Now that your freshly-turned soil has had a few days to sit and rest, it will be drier and easier to turn in the compost and fertilizer. According to the package directions, sprinkle the appropriate amount of fertilizer granules for the size of your garden, and then spread an even layer of compost over the soil. Using a pitchfork, turn the compost, fertilizer, and soil over once or twice. You don't have to go crazy, just start incorporating everything. When you've finished, water the empty garden. You want to give the whole thing a good soaking to start leeching the compost and fertilizer down into the soil.

Now that you've got your plot turned and prepared, it's time to start thinking about what plants you want to grow and what supplies you'll need to begin and maintain your garden. Below, you will find a list of the basics you might want for seeding, planting, and maintenance, some of which we already discussed in the design and tilling phase:

- seedling cells, trays, or small containers

- potting soil

- trowel

- hand rake

- weeding fork

- sturdy gardening gloves

- foam kneeler or stool

- hand pruners

- utility scissors/utility knife

- shovel/spade

- hoe/garden ax

- pitchfork

- garden rake/leaf rake

- watering can/hose

- weed bucket/basket

- trellises/stakes/veggie cages

- twine

- garden cart or wheelbarrow

- a storage bucket, chest, or shed to keep your tools safe and dry

- metal fence posts and a roll of chicken wire or snow fencing

- a notebook or journal, or software viable for journaling

There are a lot of fancy gardening gadgets available on the market, like with any hobby, but you don't need to go all out on tools. Look for bargains and end-of-season sales, and search listings on your local online buy/sell/trade groups. Even tools that have seen better days can be fixed up quickly with a good sharpening, or sanding and painting of the handle. You can keep your budget down by being a secondhand shopper and trying not to fall prey to Shiny New Tool Syndrome.

Once you've outfitted yourself with the supplies you need and want, it's time to move on to the fun part (yes, I know I keep saying every part is the fun part!), and that's choosing seeds and seedlings to populate your garden. It's also time to crack open your garden journal. Make a note of the dimensions of your brand new garden plot, and record your dates of tillage, composting, and fertilization. You can also tuck your design sketches into the journal for future reference. Now you can begin to jot down seeds and seedlings that interest you so you can look whether those varieties will be right for your garden.

Chapter 2: Choosing Seeds and Plants for Beginners and Novices

Before you get your heart set on certain plants, the first thing you need to know is your plant hardiness zone. This is the climate region in which you reside, and in the United States, they are assigned and monitored by the Department of Agriculture. When you're choosing seeds and seedlings, you're going to want to choose varieties that are indicated to grow well in your hardiness zone.

For vegetable gardening, this isn't as restrictive as it is for ornamental plants; those tend to be a little pickier about their growing conditions. The main consideration about growing zones and vegetables is the length of the growing season in your region, and being sure to choose varieties that will have enough time to grow, flower, fruit, and be harvested if you live in a zone with a short season. In warmer zones, you may be able to get two full sowings and harvests complete in one growing season. For our purposes, we're going to split the difference and use examples and information that applies to growing vegetables in a temperate zone.

So now that you know that you're going to be looking for your hardiness zone on seed packets and seedlings, what other information can we find on those tags or catalog listings? There's actually a ton of pertinent data to be found, so let's go over what you'll find:

- Seed packets will tell you EVERYTHING you need to know for a full growing season -

- Species and variety, often both the scientific and common name

- This will help you know if you've got the correct seeds or seedlings, based on what you've decided to grow. It's also a fun exercise to learn both names, you can whip out the Latin at dinner parties and impress your friends.

- Photo and description of the mature plant, including size

- This will give you an idea of what you can expect when the plant is full-grown, given that it develops properly. Some seed packets will indicate whether the variety is good for containers, as well. This generally means that the plant is compact, which means it will work in small garden spaces, too.

- *Mature qualities of the flowers, fruit, veggies, or herbs*

- By knowing what the 'finished product' will look like, you'll know what to expect as the plant is growing and will be able to judge if it's progressing correctly. And because humans eat with their eyes, it's nice to see ripe fruits and veggies so we can choose to grow what's most appealing to us to eat.

- *Hardiness zone*

- We've already discussed how important this is, but just to reiterate, you want to make sure you're putting the right plants into your garden. You'll have happier plants and a happier, less frustrated you.

- *Sowing, transplanting, and plant spacing guidelines*

- Knowing whether your seeds need to be planted indoors or directly into the garden is a valuable piece of information. You'll also need to know seed depth and spacing, and when to plant and transplant- so be sure to read this part of the packet carefully, and either write the info in your garden journal or save the packet.

- *Watering requirements*

- As a general rule, native soil vegetable gardens need approximately four inches of water a week to thrive. Seed and plant descriptions found on the packaging will tell you if plants require extra or less water for optimal growth. To be honest, you'll most often see phrasing like 'keep soil evenly moist'. The best way to do this is to just water every day unless you've had or are expecting a good soaking rain.

- Sun requirements

- Vegetables are sunseekers, and most packets will tell you that the plants require full sun. Some herbs, fruit, flowers, and other ornamentals may tolerate partial or full shade, but for the purposes of a traditional vegetable garden, expect to see 'full sun' on the packets. This is a minimum of six hours a day.

- Days to germinate and days to maturity/harvest

- This information tells you how long it will take for your plant to emerge from the seed, and then grow, mature, and bear harvestable fruit. Some varieties will produce one crop and be done, and some will continue to bear fruit until for the rest of the season. Knowing when to expect your first harvest lets you decide what varieties you'd like to plant, helps you track the progress of your vegetables as they grow, and gives you a guideline about when you can begin harvesting.

- Whether the seeds are heirloom/hybrid/GMO or non-GMO

- Here's another little botany lesson, and what you ultimately decide to plant is based on personal preference. Plants that are 'heirloom' are genetic copies of their parent plant, cultivated from the seeds that those plants produce. If you save a seed from an heirloom variety, you can sow it and get the same exact plant. 'Hybrid' varieties have been carefully crossbred by botanists or agriculturalists to exhibit the desirable traits of two or more heirloom varieties. The seeds of hybrid plants can be planted, but there is no guarantee that you will get the same plant. You may get a plant that exhibits the traits of one of its parent varieties.

- That brings us to the GMO/non-GMO discussion. GMO stands for 'genetically modified organism', which immediately turns some people off, because who wants to eat mutant food? GMOs were originally created to fill a need that hybrids and heirlooms cannot, and one of the first successful agricultural applications was to create a variety of corn that could grow in arid conditions, therefore being able to create a larger food supply in destitute African desert nations. The first commercially available GMO food in the United States was the Flavr Savr tomato, which you might remember if you're of a certain age. It was genetically modified to delay ripening off the vine, so it stayed fresher longer on supermarket shelves. Today, GMOs get a bad rap not because of the science, but because of their applications and the questionable motives of some of the companies that produce them. Whether you choose GMO or non-GMO seeds and plants is entirely up to you, but at least you're now armed with the background to make that choice or do more research, should you choose.

- *Name of the producer or seed company with contact information*

- On seed packets and plant tags, you will see the name of the company, an address, and more often than not these days, a website or social media icon. In the United States, the USDA requires that producers provide this contact information, and many other countries similarly regulate their agricultural products. One great thing to benefit gardening hobbyists in the internet era is instant customer service and reliable online product information. You can hop on your computer, find the seed company's website and either find the answer to your questions or speak directly with someone who can. It's a blessing to beginning gardeners to have a direct route to speak with the company that produced their garden seeds.

- The 'Best by' date

- This is the company's date of viability, and it's usually two to three years. While many seeds, if stored properly, will be viable for much longer, seed companies want you to use them by the date stamped for best results. If you've got some old seeds laying around and you're not sure if they're still good, take a couple, place them in a damp paper towel for a couple of days, and see if they begin to open up and germinate. If they do, seeds are still good. If not, toss them in the compost heap. Sometimes instead of a date, the packet will say 'packed for 20xx growing season' or something similar. That still lets you know how old the seeds are.

- The quantity

- Seeds, like people, come in all shapes and sizes. You may have two packets of seeds that are the same exact dimensions, but open them up, and you'll find several hundred carrot seeds in one and two dozen pumpkin seeds in the other. The most common measurements for seed quantities in a packet are the number of seeds, by the ounce, or by the gram. Pay attention to this, especially if you're purchasing seeds online. Miss a unit of measurement, and you may be keeping your entire neighborhood in tomatoes for the next twenty years.

As you can see, seed packets, plant tags, and catalog listings can pack a lot of information into a small space, but that's great! These tags give you everything you need to know about that specific plant. There aren't too many other products on the market today aside from a seed packet that can give the consumer that much information in such a small space.

There can be so many varieties to choose from that it's easy for beginning gardeners to get overwhelmed. Let's take a look at some common kitchen garden items that are easy to grow and maintain, and will give you a great harvest from year one.

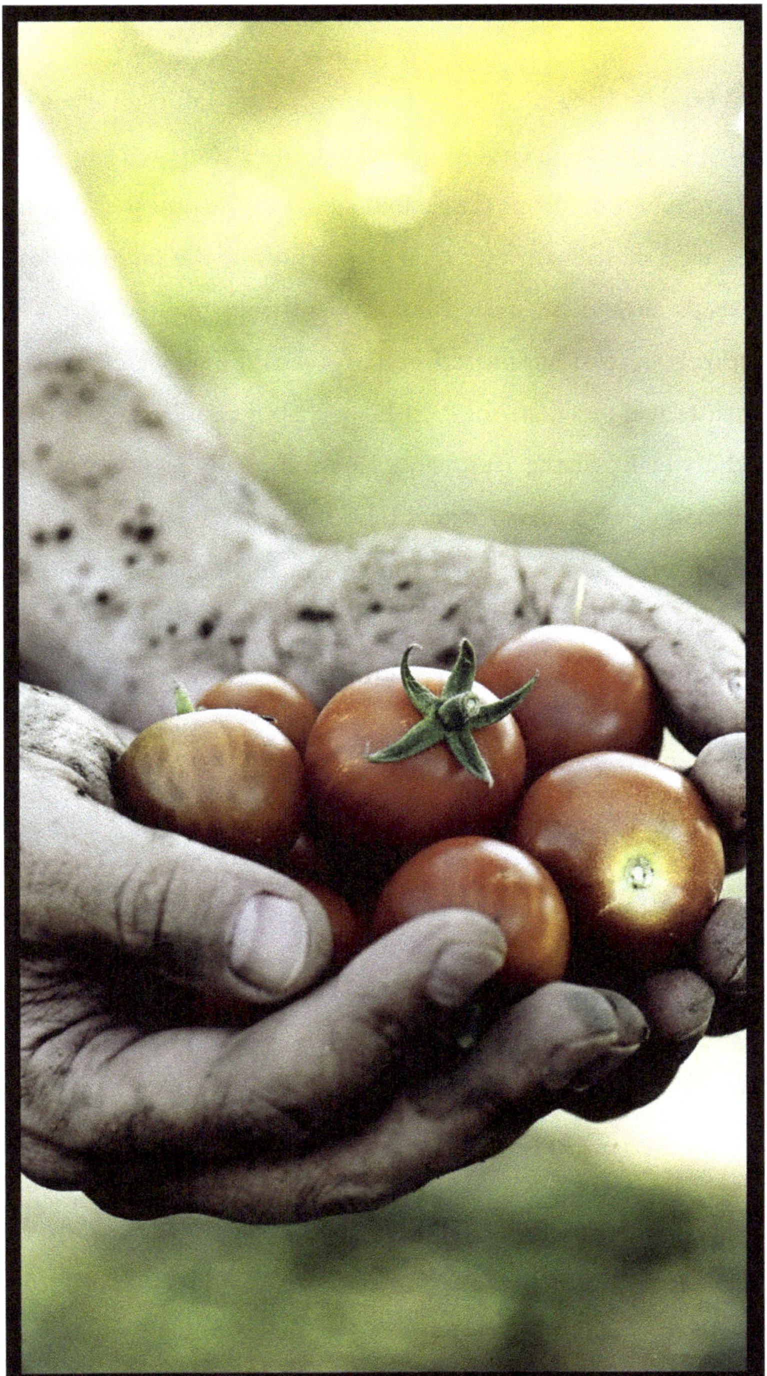

- *Tomatoes*

- You'd be hard-pressed to find anyone who doesn't like tomatoes in some shape or form- even if it's just in ketchup. There is a tomato variety out there for everyone, from classic beefsteaks to the tiniest cherry tomato. Tomatoes are a classic summer fruit (yes, fruit!) of the nightshade family. Tomato seeds should be started indoors and transplanted to the garden once the threat of frost has passed. When given enough water and sunlight, tomato plants will grow strong and healthy, but require stakes or caging because the weight of the fruit can cause the plants to sag and tip over. Once a plant begins to fruit, it will bear harvest until late in the season. Even the end of the year's green tomatoes can be taken in and used to make fried green tomatoes and salsa verde. Tomato plants are very heavy feeders and may require fertilization during the growing season to stay on track.

- *Beans/peas*

- Beans and peas, which come in many varieties, are the givers of the kitchen garden. They belong to a family of leguminous plants, and legumes have a really cool function- they can put nitrogen back into the soil as they grow! This is due to a symbiotic relationship formed with a soil microorganism known as rhizobia. The roots of the legumes exude excess sugar from photosynthesis, which the rhizobia latch onto in droves. Their nitrogen waste replenishes the soil around the roots of the legumes. Beans and peas come in both bush and vining varieties, require regular water and light, and grow quickly when sown directly in the garden in early spring. Harvest them early and often for a continuous flow of fresh beans and peas, which can be eaten raw or cooked, or dried, frozen, and canned to preserve them for the off-season. Vining varieties should be trellised or staked to provide support for climbing.

- *Cucumbers*

- Cucumbers belong to the cucurbit family, and they also come in a lot of neat varieties. The classic cucumber is a long, dark green fruit, but smaller varieties and color variations are common these days. The cucumber hasn't changed much over the centuries- Roman records show that cucumbers were grown for the ruling class during the height of the empire, and seem to indicate that the plant originated in the region of what's now modern Armenia. Cucumbers aren't fussy plants and can be grown in small spaces in bush varieties. Vining varieties can be trellised to grow in small spaces or protect the plants from mildew from laying on the damp ground. Cucumbers will continue to produce as long as they are harvested, so be prepared to eat a lot of salad and make a lot of pickles if you grow a lot of cucumber plants.

- *Peppers*

- Peppers also belong to the nightshade family, and you don't have a be a fan of spicy food to grow and enjoy peppers. There are plenty of sweet varieties, the classic bell pepper of course, plus many others. Jalapenos, poblanos, and banana peppers are also fun and easy to grow. Like their cousins the tomatoes, peppers also succeed best when sown indoors and transplanted after the last frost. Cool fact about peppers- the rainbow of colors is not indicative of different varieties! Those come from when you harvest the fruit. A pepper is technically ripe when it is of size and becomes a shiny green, but as you leave it on the vine, it will begin to change color from green, to yellow, to orange, to red. Peppers can be eaten fresh raw or cooked, or prepped and frozen for the off-season.

- *Lettuces/greens*

- Looseleaf lettuce and salad greens are a great place to start growing your own leafy vegetables, and they can be among the easiest and the most frustrating to grow- but the challenge is definitely worth it! The great thing about greens like lettuce, kale, and spinach is that they are cold-hardy, which means you can plant them outdoors early and enjoy them all season long. The challenge in growing greens comes from keeping woodland creatures like chipmunks, squirrels, and groundhogs away from them while they grow. Consider making little screens or putting netting over your budding greens to keep them safe while they are maturing. You don't want your tender young plants to become a rodent buffet.

- *Squash*

- Once you become a gardener, you'll get in on the joke about not being able to give away mid-summer zucchini. Squashes are also in the cucurbit family and they can be prolific producers. They're also heavy feeders that can rapidly deplete your garden soil, so they should be planted sparingly. One or two plants of each variety will provide you with plenty of squash for the whole season. Aside from the ubiquitous zucchini, you can plant other summer varieties, like the classic yellow, or try your hand at winter varieties like butternut, acorn, or spaghetti squash. Squash does best when sown directly into the garden after the last frost.

- Herbs

- A kitchen garden isn't complete without some herbs to cook with your fresh vegetables. Many common herbs are perennial, which means you can plant them and enjoy them for years to come. They also make great companion plants for warding off unwanted insect pests from invading your vegetable plants. Consider planting your own basil, dill, thyme, sage, rosemary, oregano, or mint- you can pick and choose based on what you like to cook. When you can step outside and snip off your own fresh herbs for dinner or dry bunches of herbs to enjoy in your winter soups and stews, it's a next-level experience for your taste buds. Many herbs can be grown from seed, starting them indoors or outdoors depending on your frost conditions.

- Alliums

- The most commonly-grown home garden alliums are onions and garlic, and since those are staples for many common recipes, wouldn't it be nice to cultivate your own? You should plan ahead to plant onions and garlic, as they are generally planted in the fall and harvested the following spring and summer. Alliums aren't usually sown from seen in a kitchen garden, but instead are grown from cuts or 'sets' from the previous year's harvest. If you decide to grow alliums, you can check out the different varieties and their characteristics by doing a little research in the summer, then choosing, purchasing, and planting your sets in the fall.

- Strawberries

- Nothing says summer like fresh berries, and strawberry plants make awesome companions for vegetables in your garden by providing ground cover and drawing pollinating insects. You can even trellis strawberry vines to grow up and around the rest of your garden, so they don't take up too much space. Strawberries are best grown from commercially-available seedlings, but once you've established them in your garden, you'll be able to enjoy them for several years.

- *Melons*

- Melons are the most popular fruit in the cucurbit family because they are fabulous for summer eating and don't require much maintenance to grow. Common melon varieties are watermelon, cantaloupe, and honeydew, and there are small versions available if you don't want the vines taking over the whole garden. Most melon species should be planted outdoors in the mid-spring. They are not as heavy of feeders as their cousins the squash and cucumber, but they can be thirsty- melons have extremely high water content and require a lot of water to develop properly.

These are just some of the wide variety of plants you can grow in a traditional vegetable garden, but these are among the best 'starter' species. Once you've become more confident in your abilities, you can start to branch out into species like brassicas like broccoli and cauliflower, tubers like potatoes, and root vegetables like beets and carrots.

Chapter 3: Planting and Maintenance

Whether you decide to start your garden from seed or from nursery stock seedlings, once you've chosen what to grow, it's time to get planting. If you're going to be starting from seed, indoor or outdoor, you'll need to pay close attention to the sowing instructions on your seed packets. The spacing and depth guidelines will help you get your seeds off to the best possible start.

Sowing and Transplanting

For indoor sowing, make sure that you've got a good light source for your seedlings when they begin to sprout. If you don't have a viable window that you can set your pots or seedling cells near, you can invest in an inexpensive grow light. While some units can run on the pricier side, there are small portable units to be found for less than $50, and they can prove invaluable in your gardening endeavors. You want your newly germinated plants to get off to the best start with enough warmth, light, and water.

When it's time to transplant your seedlings, either those you've grown or those you've purchased, you also need to follow the spacing guidelines. You can use a trowel to make holes slightly deeper and wider than the seedling pots. To remove your seedlings, gently squeeze the pots to loosen the soil, and grasp the seedling by the base of the stem- never by the leaves. Settle your seedling into its new home and gently but firmly tamp the soil around the plant. Once you've got all your seedlings installed, give them a good drink, and leave them be to nestle into their spot. The last thing to do after you're all planted is to pound your fence posts into the ground and wrap your chicken wire or snow fence protection around the garden. If that's not your style, you can certainly choose to put up something more aesthetically pleasing. Be sure to mark all your planting and transplanting dates in your garden journal.

If you're unsure how to arrange the plants in your garden, it's a good rule to keep your heavy feeders apart from each other, as well as plants from the same family. You can organize your varieties by placing a legume in between your heavy feeders like peppers, tomatoes, and squash. Intermix your herbs and lighter-feeding cucurbits, and you'll have a give-and-take relationship among your plants and your soil that will avoid depleting your soil of nutrients too quickly.

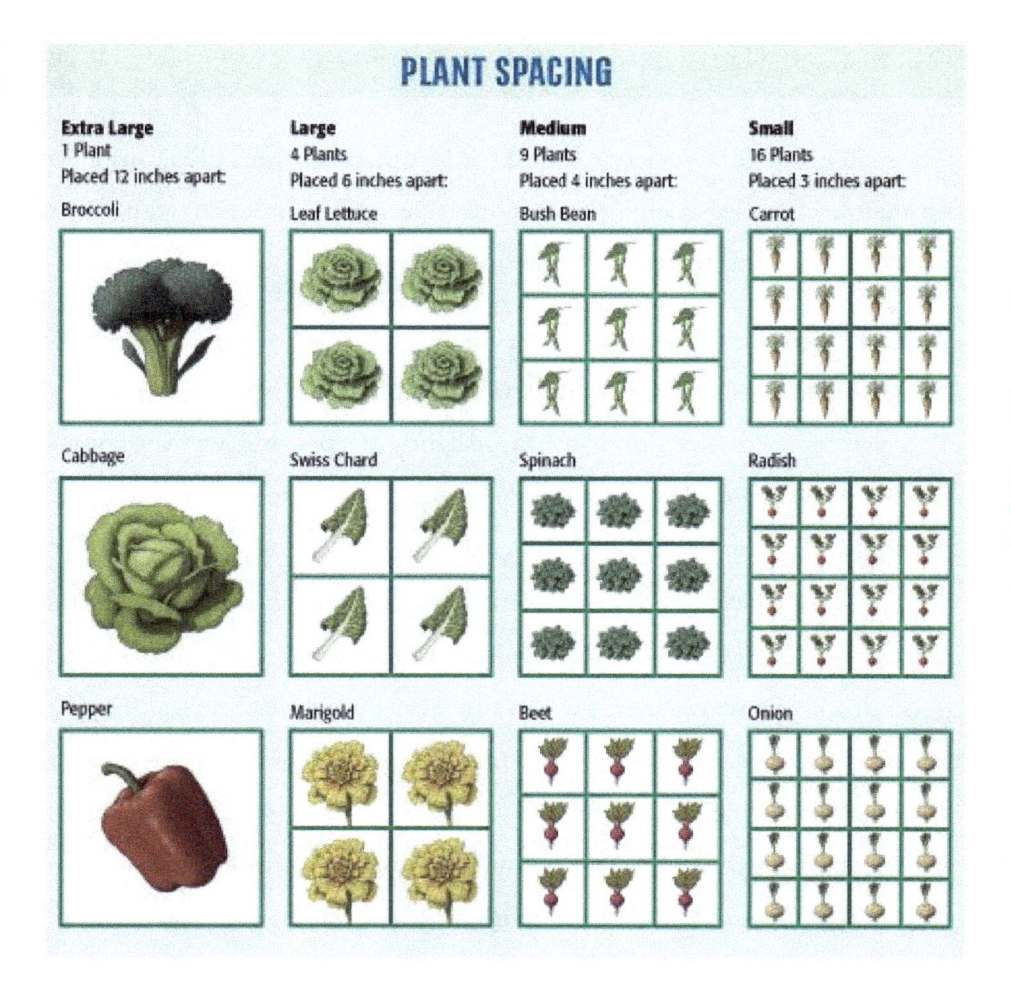

PLANT SPACING

Extra Large	Large	Medium	Small
1 Plant	4 Plants	9 Plants	16 Plants
Placed 12 inches apart:	Placed 6 inches apart:	Placed 4 inches apart:	Placed 3 inches apart:
Broccoli	Leaf Lettuce	Bush Bean	Carrot
Cabbage	Swiss Chard	Spinach	Radish
Pepper	Marigold	Beet	Onion

- Keep plant size in mind when you are planning and planting, and use recommended guidelines -

Watering

You should make a point to water your garden every day unless of course, you are expecting heavy rain. Tomatoes are a great non-scientific measure of whether or not you are overwatering- the fruit will begin to split if the tomato plants are overhydrated. Dial it back a little bit with the watering. Remember, you can always water again, but you can't mop up excess moisture if you've overwatered. If you like, you can easily install a self-watering system like drip irrigation or soaker hoses; these come in fairly inexpensive DIY kits at most garden centers or home improvement stores. Add a timer to your hose spigot, and you've got a set-it-and-forget-it watering method.

Some people love to water, whether by hose or watering can, and that's great, too. One important thing to remember when your watering your garden is to water the roots, not the leaves. Leaves aren't the part of the plant that takes in water, and wet leaves can make your plants susceptible to sunscald and pathogens. If you can, always try to water in the morning so the soil has time to dry over the course of the day. You should also avoid using sprinklers for vegetable gardens because they water unevenly and cannot target the roots. Save sprinklers for the front lawn and for the kids and grandkids to run through.

Fertilizing and Composting

Fertilizer and compost can help you get your vegetable garden through a long growing season without stripping all the nutrients from your soil.

Remember what we talked about earlier, though. Fertilizer can be toxic to your plants in too-large doses because more is not always better. Plants can suffer from nitrogen, phosphorus, and potassium toxicity just as easily as they can suffer from a deficiency. The best time to fertilize your garden is at the halfway mark of the growing season. You can use the same granulated fertilizer you used to prepare your soil for planting. Following package directions, sprinkle an appropriate amount in the garden before your morning watering. When you fertilize mid-season, it's called 'side-dressing' your plants.

Compost is always welcome. As noted before, you cannot put too much organic matter into your soil. If you have the space to start your own compost heap, that's even better, but if not, bagged or bulk will always be preferable to nothing. You don't need much space to make compost for yourself, and pallet piles are all the rage. All you need is 8 metal fenceposts, four pallets, and your pitchfork. Arrange the fenceposts in a square, so that the pallets will all slide down over the posts to create the walls. Then put all your plant waste, grass clippings, household foods scraps (no fats or meats), eggshells, used paper towels, shredded newspapers, etc. inside the square. Wet it down with the hose and cover it with a tarp- bonus points for bungee-cording it to the pallets. You can remove one pallet to use your pitchfork to turn the pile once a week or so. The material on the bottom of the pile will begin to decompose. Just keep the pile warm, wet, and turned, and you will have your own rich, crumbly compost (known as humus) to enrich your garden in no time.

Building a layered compost heap
1. Build your compost in thin layers (3 to 10cm).
2. Alternate high nitrogen (e.g. food scraps) and low nitrogen (e.g. dry leaves) layers.
3. Aim for a ratio of 3 buckets low nitrogen to 1 bucket high nitrogen.
4. Use a diversity of materials.

This diagram is an example of the different layers. Alternating kitchen and garden waste layers with an occasional layer of manure works well.

	LAYER OF HESSIAN TO RETAIN HEAT AND MOISTURE
	LOW NITROGEN ·············· STRAW AND WATER
	HIGH NITROGEN ·············· KITCHEN WASTE
	WATER
	LOW NITROGEN ·············· GARDEN WASTE
	HIGH NITROGEN ·············· MANURE
	LOW NITROGEN ·············· COARSE PRUNINGS
	HIGH NITROGEN ·············· GRASS CLIPPINGS/PAPER
	LOW NITROGEN ·············· STRAW OR DRY LEAVES
	HIGH NITROGEN ·············· SOFT PRUNINGS
	LOW NITROGEN ·············· COARSE PRUNINGS
	LOOSELY FORKED SOIL BASE

- Composting is simple when you know what to include! -

Troubleshooting Weeds, Pests, and Disease

Now to the nuisance stuff- weeds, pests, and pathogens. The sad truth is, no matter how closely you follow best gardening practices, these three will find a way to crash the party. The next few paragraphs will take you through some ways to avoid, diagnose, and alleviate problems that may arise, and hopefully, you won't have to deal with too many of these issues as your build and cultivate your new garden.

Weeds are the 'easiest' problem to deal with, so let's start there. You can avoid many weed issues by mulching your vegetable garden, but we don't mean using regular wood chips as you might use in an ornamental garden. These can be hard to turn into the soil and may not break down at the end of the season, so they are ideal. Some hardwood chips can also leach tannins into your soil, which will disrupt the pH. Some of the best natural mulches for veggie gardens are straw (not hay!), grass clippings, and weed cloth, which is exactly what it sounds like- rolls of material you can cut and lay among your plants to cover the bare soil between them. Mulch has the added benefit of insulating the soil and helping it retain water.

When you manually weed, you can compost any material that hasn't yet gone to seed. Anything that has bolted should go in the trash- you don't want any of the seedy stuff headed back into the garden as compost the next year. You should avoid at all measures using any herbicides in your vegetable garden. Most commercially-available weed killers are broad-spectrum for broad-leaf plants, which means they will take down anything in their path. These chemicals cannot differentiate between a dandelion and a daffodil, and have no business being sprayed near your growing food. Instead, invest in mulch and a good weeding hoe. Your garden will be much happier for it.

Pests are an unfortunate part of garden life, too. You should become familiar with some of the most common vegetable garden pests, so you're not surprised when you see them. Tiny green aphids, globular slugs, and Japanese beetles are three of the most-seen garden nuisances because they aren't too picky about what you are growing. Some common species-specific pests you might see are squash vine borers, which adore cucurbits, and tomato hornworms, which as you can probably guess, love tomatoes.

Here's the thing about garden insects, though- some are super helpful. Ladybugs and praying mantises eat harmful insects, and bees, butterflies, and dragonflies are fantastic pollinators. If you see an insect in your garden, and you're not sure what it is, DON'T touch it. You don't know if it's something that can bite or cause skin irritation. If you can, take a photo. Look it up online, using the most specific search terms you can, i.e. "one-inch long black and silver beetle with large antennae found in the northern part of X county in Y state".

Once you've identified what's eating your garden, you can take appropriate steps to alleviate the problem. Chemical pesticides should be used sparingly in vegetable gardens, and you can opt for horticultural soaps and oils that do the trick nicely. You can also try pheromone traps, like the bags you'll often see hung to attract and trap Japanese beetles. Here's a neat tip to project whether or not you'll have a Japanese beetle problem in the summertime- watch how many corvids (crows, starlings, and/or ravens) come and eat from your lawn in the fall and spring. These birds LOVE grubs, which are the Japanese beetle larvae that were laid in the ground. Lots of corvids feasting= lots of grubs in your lawn = banner year for beetles.

Birds are a great way to manage insects in your garden, by the way. If you can, put birdhouses, feeders, and birdbaths around your property to attract the omnivorous local species that can help control the insect population in your garden. This type of holistic approach of using non-chemical and biological controls is called integrated pest management, or IPM. IPM is becoming a popular method for both home gardeners and large-scale growers to reduce the use of harsh pesticides that can permanently damage the ecosystem.

What is IPM?

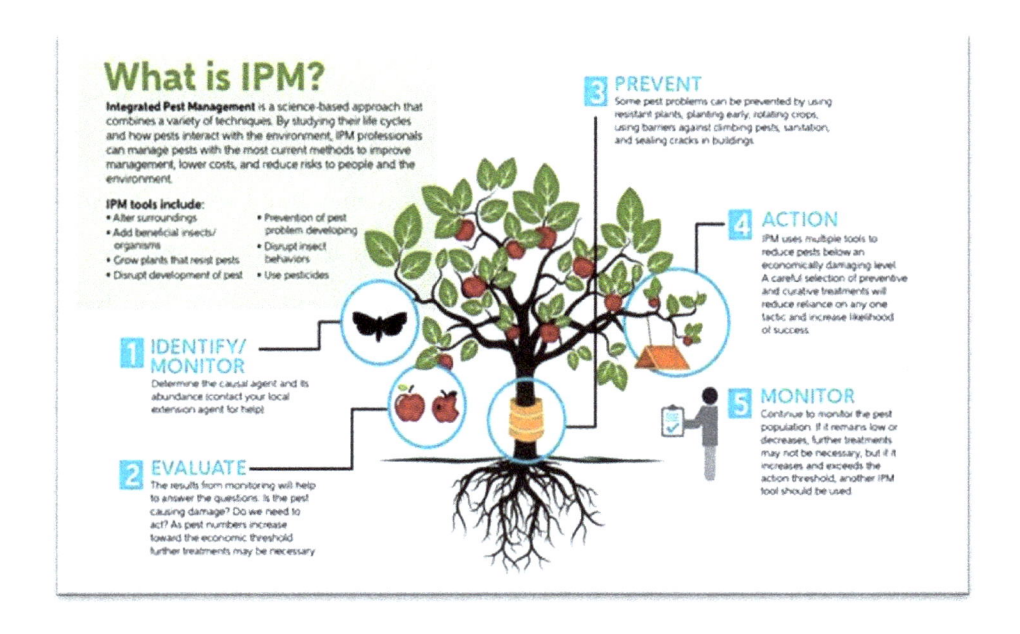

Integrated Pest Management is a science-based approach that combines a variety of techniques. By studying their life cycles and how pests interact with the environment, IPM professionals can manage pests with the most current methods to improve management, lower costs, and reduce risks to people and the environment.

IPM tools include:
- Alter surroundings
- Add beneficial insects/organisms
- Grow plants that resist pests
- Disrupt development of pest
- Prevention of pest problem developing
- Disrupt insect behaviors
- Use pesticides

1 IDENTIFY/MONITOR
Determine the causal agent and its abundance (contact your local extension agent for help)

2 EVALUATE
The results from monitoring will help to answer the questions: Is the pest causing damage? Do we need to act? As pest numbers increase toward the economic threshold further treatments may be necessary

3 PREVENT
Some pest problems can be prevented by using resistant plants, planting early, rotating crops, using barriers against climbing pests, sanitation, and sealing cracks in buildings.

4 ACTION
IPM uses multiple tools to reduce pests below an economically damaging level. A careful selection of preventive and curative treatments will reduce reliance on any one tactic and increase likelihood of success.

5 MONITOR
Continue to monitor the pest population. If it remains low or decreases, further treatments may not be necessary, but if it increases and exceeds the action threshold, another IPM tool should be used.

- IPM can help you avoid major pest concerns in your garden -

Pathogens are another common concern for vegetable gardeners. While you can do your best to avoid creating conditions for disease to thrive, they can sometimes pop up despite your efforts. Plants can be susceptible to viruses, bacteria, fungi, and these conditions most often manifest their symptoms on the leaves of the affected plant material. Some of the most common pathogen issues in vegetable gardens are powdery and/or downy mildew, early and late blight, mosaic virus and root rot, and leaf spot or leaf curl. When you see something 'off' about your plant, you should take a photo of the affected part of the plant and do a quick online search to see what's going on.

Sometimes, you'll be able to salvage the whole plant; cucurbits are prone to powdery mildew, which makes the leaves look terrible, but doesn't affect the flowering or fruiting function of the plant. One thing you should do with all diseased plants is cut away the affected portion and dispose of it in the trash, not the compost- these aren't the good microorganisms you want in your compost heap. Other times, you may need to call it a loss, remove the plant, and move on. As long as you follow best practices by watering your roots, spacing your plants for adequate airflow, and being vigilant about identifying and removing diseased plant tissue in a timely fashion, you should be able to avoid or alleviate any pathogen issues with ease.

Chapter 4: Harvesting and Preserving

Harvesting your garden is one of the greatest joys in the whole hobby, and when's time to begin reaping what you've sown, there are a few tips you should follow. As a general rule, you should harvest in the morning before your daily watering. An old kitchen colander is one of the best harvest baskets because you can carry it inside, pop it right in the sink, and wash your morning haul. Harvesting in the morning takes the fresh fruit and veggies off the plant before the heat of the day.

You should take care to never tear the plants when you harvest. For things like beans and peas, remove the pods from the vine not at the top of the pod itself, but at the little joint where the pod's stem meets the vine. For cucurbits, try to remove the whole stem from the vine when you pick- you can do this by twisting or cutting the stem at the joint with the main stem. Tomatoes will tell you when they're ready to go by essentially falling off the vine into your hand. Bumping into a heavily-laded cherry tomato plant will send a cascade of fruit falling to the ground, and you scrambling to pick up every last juicy treasure.

You can pick some items right before they ripen, too- cucumbers and green beans are less pithy and much sweeter when picked on the small size. Smaller cukes are crunchier for making pickles, too. You should also pick squash and zucchini before they get to comical proportions, or else you'll have monsters on your hands with large, unpalatable seeds. After harvest, give your produce a good cold water rinse and pat or lay out to dry. You don't need to immediately refrigerate most fresh veggies. If anything gets bruised or split, make a point to either eat or cook and store those items the same day, or place them in the compost. Broken skin on vegetables can invite bacteria. You can make notes in your garden journal about what grew nicely and provided an abundant harvest, and what varieties didn't fare as well. It will help you make decisions about what to plant in the future.

The great thing about having a kitchen garden is having fresh produce right outside your backdoor, but you can also save your harvest for enjoyment in the off-season. The easiest way to do this is by freezing your fruits, and blanching and freezing your vegetables. Berries freeze very well, and only need a good wash before being bagged and labeled. Frozen berries can be used later for pies and smoothies, or eaten as is as a cool treat. Frozen vegetables prepared at home will outshine those you can find at the grocery store. You can freeze beans, peas, sliced peppers, sliced and shredded zucchini, and much more.

For those who are feeling more ambitious, canning is a great way to fill your pantry with pickles, jams, and salsas for the coming off-season. Most items from your garden can be canned with the simple water bath method, which requires very little equipment, and you can find cookbooks and online recipes with exceptional instructions and traditional and contemporary flavor combinations to tempt any palate. Some people like to vacuum seal their prepared foods as well, and inexpensive countertop systems have made this a great option for people who don't have much freezer space but are shy of the chore of canning.

Drying and dry storage are also among your choices for extending the life of your harvest. Beans and peas can be dried in or out of their pods and make terrific additions to soups and stews. You can also dry bunches of herbs, that can then be crumbled and stored in airtight glass or plastic jars to use for seasoning in all your favorite dishes. If you've grown onions and garlic, you can cure and store these items to enjoy for months to come! By employing any of these preservation methods, you'll be eating healthy, homegrown food long after the growing season has ended.

Chapter 5: Preparing for the Future

Once your garden has exhausted itself, it's time to think about putting it to bed for the winter. You should clean out all spent plant material and compost it, given that it is free of disease. Make sure you remove all roots, vines, and fallen fruits and vegetables, lest you accidentally seed something where it shouldn't be seeded. Spread a layer of compost in the garden and give it a quick turn. Lastly, you should put an organic ground cover on your soil- either by planting a winter cover like clover, rye, or vetch, or by spreading a layer of straw (not hay, which has seeds!). When spring comes, you can turn what left of that layer or cover crop right into the garden to improve your soil organic matter.

When you plan your garden for the following year, you should practice crop rotation to make sure that you don't put heavy feeders back into the same space which will strip the soil. Move everything one spot to the right, if you grow in rows, or one spot clockwise if you grow in squares or circles. You can do this every year so that your legumes have a chance to enrich the soil before a moderate or heavy feeder moves back in. Come spring, you'll want to turn the garden, fertilize, and compost, like you did when preparing your garden for the first time.

The off-season doesn't have to mean you stop being a gardener! Take time over the winter to browse seed catalogs, read up on new gardening gadgets, and browse online gardening resource sites. Everyone who gardens eventually finds a niche, whether it's playing around with cucumber varieties to make the best pickles, or developing herb mixes to gift to family and friends. Look for an online community of like-minded gardeners to exchange ideas with- you'll find your tribe whether you're interested in growing award-winning tomatoes or obscure bean varieties, or learning more about soil science. Take notes and jot ideas and sketches in your gardening journal, you'll be amazed at how the ideas start flowing when you start researching.

Most of all you should enjoy the journey. You've taken a bare plot of ground and turned it into a food-producing paradise, and done it in the span of just a few months! You should congratulate yourself on your success. You can now take what you've learned and become a stronger, more capable gardener with every growing season.

PART III

Raised Bed Gardening

Are you thinking about building a garden, but you don't have the time, space, or inclination to till up a patch of open ground? Maybe you've always wanted to grow fresh, healthy food or beautiful flowers at home, but you've got mobility concerns or physical limitations that won't allow you to bend and twist in a traditional garden. Or maybe you live on a rental property where you want a garden, but it can't be a permanent installation. No worries! You can be a gardener, and you can do it with raised beds!

Raised bed gardens are a great way to address any of these restrictions, and with the development of compact varieties of many common garden plants, you can grow almost anything you'd normally see in a traditional native soil garden. By definition, a raised bed garden is any garden that utilizes materials to create a planting space that sits above the elevation of the ground, and the only limit on what you can build is your imagination. Raised bed gardens can be built from a variety of materials, eliminate much of the guesswork of soil quality, and will provide a rewarding hobby you'll enjoy for years to come.

Chapter 1: Designing and Building a Raised Bed Garden

Before you can plant a raised bed garden, you've got to build a raised bed garden, and that means finding a great spot and selecting a design. Since most flowers and herbs require about six hours of sun a day, and six hours is the absolute minimum of sunlight for vegetables, the first thing you need to consider is finding a well-lit spot for your garden beds.

The next thing that comes into play when choosing a site for a raised bed garden is the slope. You want to choose a site that is level, can be leveled, or can be terraced. While a gentle slope is good for drainage, a more dramatic slope can be an invitation for erosion and runoff, even in a contained bed. You should also take into consideration any trees that are nearby. While they might not cast shade now, is there a possibility that they may grow and block the sun from your garden in the future?

A third consideration for raised bed gardens is their proximity to a water source. Because raised beds tend to be warmer and have more drainage than a traditional in-ground garden, they can be thirsty. You want to make sure you've got convenient access to water so you don't have to travel far to fulfill your garden's watering needs. It's easy to water when the hose is right there, and it's easier to put off when you don't feel like dragging that hose all over every day.

Once you've chosen a site that best fits your criteria, it's time to think about your design. Creating a raised bed garden is fun because the possibilities are endless. Some things to think about when you're designing your garden are bed dimensions, bed height, bed arrangement, and what materials you'd like to use. Keep in mind that you can build raised beds from hardwood, masonry block, PVC vinyl, or composite materials. You can also take advantage of the many commercially available pre-fabricated kits to create your garden design.

When considering dimensions, keep in mind that if you are gardening for food, you need about 200 sq. feet per person in the household throughout a growing season. If you're going to garden for flowers or a mix of food and flowers, you can go crazy with whatever dimensions you'd like. The key is to make sure that your beds are deep enough to support the roots of your growing plants, with 12 inches being a standard minimum. Using good old-fashioned pencil and paper, sketch a few ideas. You can go with a set of squares arranged in a quadrant, or some rectangles set in rows like a traditional garden. Think about using shapes like triangles to fill in corner spaces, or building adjacent boxes in increasing heights like steps to create a more interesting space and accommodate plants of different heights.

Material & DESIGN OPTIONS

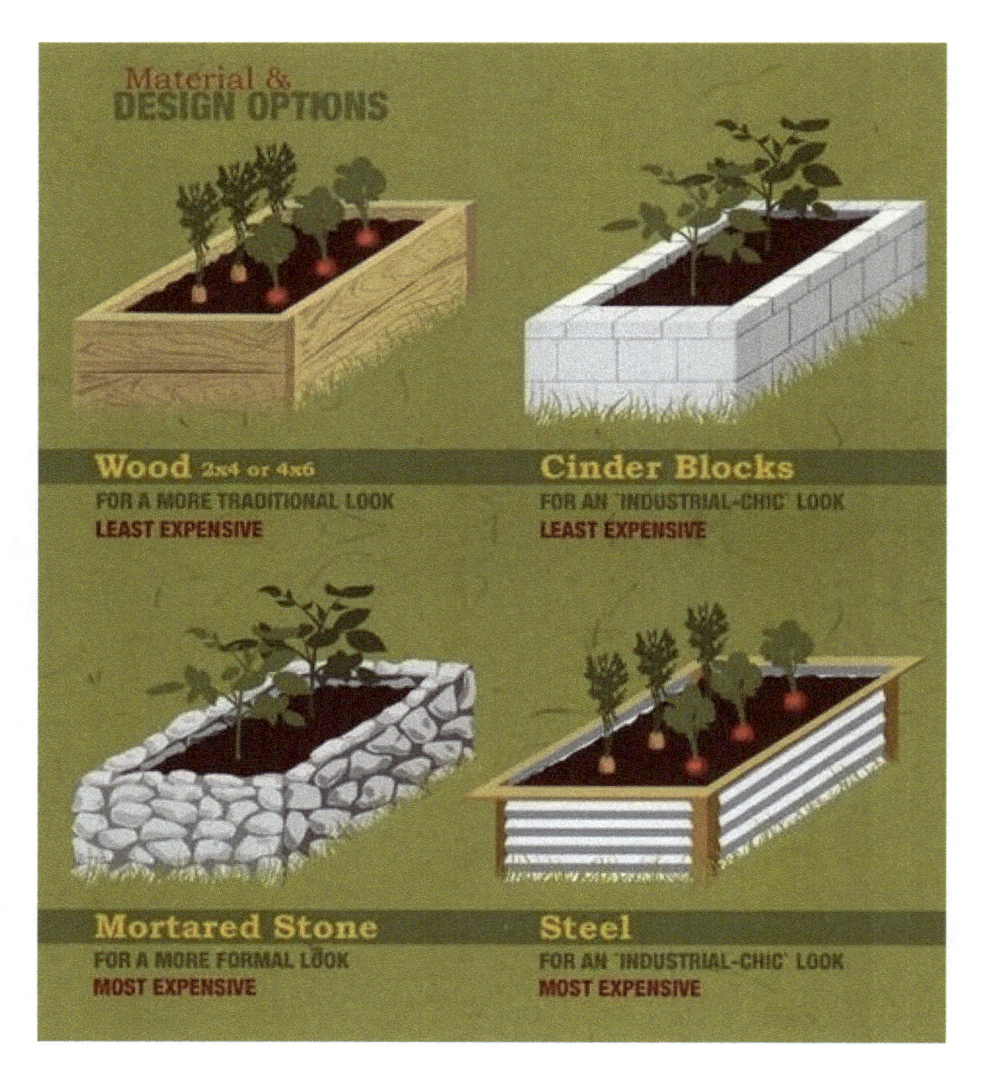

Wood 2x4 or 4x6
FOR A MORE TRADITIONAL LOOK
LEAST EXPENSIVE

Cinder Blocks
FOR AN 'INDUSTRIAL-CHIC' LOOK
LEAST EXPENSIVE

Mortared Stone
FOR A MORE FORMAL LOOK
MOST EXPENSIVE

Steel
FOR AN 'INDUSTRIAL-CHIC' LOOK
MOST EXPENSIVE

- Raised beds can be designed in a variety of shapes, sizes, and materials -

If you've got mobility issues, you might want to think about raised growing tables rather than beds. These can be constructed from wood, or you can purchase them in composite vinyl. These are large troughs on legs or stilts, often with locking casters to roll them into different positions. The troughs have drainage built into the underside, and are generally deep enough for many common garden plants, although you might be hard-pressed to grow root vegetables in them. Raised tables are a great way to make gardening adaptive to people of any physical capability. You should also think about what else you'd like to have in your garden to make it an inviting space. You can hang bird feeders, a table and chairs, or some whimsical gnomes. You want to brainstorm how to make your garden a place you'll enjoy working and relaxing in.

If you are going to be building your own beds, then you should look into repurposed materials. Contact any friends in the construction trades, and see if they have any leftovers you can have for cheap or the price of hauling. You can also look at your local online buy/sell/trade groups and see if anyone is selling or giving away construction materials. Sometimes, you can get discount 'cut ends' from the sawmill or lumberyard- these are pieces cut from larger planks that the contractor didn't need. The only material to avoid is pressure-treated lumber, it can leach the treatment chemicals into your garden soil, which can affect the nutrient balance and damage plant roots. Be resourceful and you'll be able to find a plethora of materials for the taking without spending a ton of money.

When it's time to build, you should make sure you've got all your materials and tools in place. Consider everything you need to measure, cut, and assemble your beds, including hardware for wooden, vinyl, and composite beds, and mortar for masonry block beds. Keep in mind the old adage to measure twice and cut once to avoid waste. If you've got a square and a level on hand, they can be invaluable tools for making sure you're on the right track for having nice, even beds. Of course, if you're handy enough to be building your own raised beds, you already know all these things!

If you've purchased pre-fabricated garden kits, make sure you open the packages, look at the directions and material lists, and ensured that all the parts and hardware are there BEFORE you start putting anything together. These kits don't generally require too many tools for assembly, but you'll probably need a screwdriver or power drill with screwdriver bits, or a wrench or rachet if it assembles with nuts and bolts. If anything is missing, contact the manufacturer to let them know. You will also want to have a level on hand for when you set your new bed into place. You should stow any paperwork, including assembly instructions, in a file for safekeeping. You may need to replace worn-out hardware in the future, and having the original paperwork will save you time and potentially money when purchasing new fasteners.

Once your beds are built, the next thing you are going to need is soil. In a traditional garden, you have to work with what you've got, amending as you go to reach optimal conditions. In a raised bed garden, you get to choose the soil and avoid some of that preparation. If you can, you should have your soil delivered in bulk from a reputable supplier. When you order, you should specify that you are going to be using the soil for gardening and that you want screened topsoil, preferably pH tested and irradiated for weeds. You want to make sure you are getting quality for your money. The supplier can help you calculate how much soil you need based on the dimensions of your beds.

When your soil is delivered, try to have it dumped as close to your beds as possible, without ruining your lawn, of course. If you can't install all your soil in one day, make sure to cover the pile with a tarp or sheet of plastic. You don't want a sudden overnight rain shower to wash all your soil away!

If you cannot get soil delivered in bulk, you can purchase bagged topsoil, but be careful about the quality. Bagged topsoil can be heavy, wet, and prone to mold. If you do go with a bagged option, make sure you fill your beds slowly, giving time between layers for the previous layer to dry out a bit before you continue.

With your beds full of soil, it's time to make them fertile! Before you plant anything, you need to add some organic matter to perk up the soil organisms that just found themselves unceremoniously dumped in a new home. You can spread a layer of compost on your beds, and turn it into the soil, or you can lay some straw (not hay!) on top of the soil and give it a good wet-down. Water the straw layer every day for a few days to encourage decomposition, and then turn it into the soil to continue to break down. You can never, repeat, never, have enough organic matter in your garden soil. Finish your soil preparation by applying some granulated or liquid all-purpose fertilizer, being sure to closely follow the packaging instructions. If you've got any questions about this, contact your local Farm Bureau or Cooperative Extension for guidance before you apply anything.

Chapter 2: Choosing Plant Varieties That Thrive in Raised Beds

When you are selecting plants for raised beds, you want to make sure that you choose varieties that are compatible with growing both outside of native soil and in a compact area. Whether you are planting vegetables, flowers, herbs, or a combination of the three, you want to look for plants that are labeled as compact, dwarf, or 'good for containers', before you read anything else on the plant tag. Yes, the plant tag. When growing in raised beds, at least for the first year until you build up your soil, it's best to start with seedlings and plant starts rather than from seed. It will make your transition from having no garden to having a freshly-built raised bed garden easier. Skipping the home-sowing step will let you go from zero to planting in no time, and with the limited space available in a raised bed garden, transplanting nursery stock or seedlings takes the guesswork out of planting.

Let's takes a look at the information you will find on a plant tag, so you know what to look for to choose what's right for your garden:

- *Hardiness zone*

    - The hardiness zone, or growing zone, is the region or subregion that is optimal for each type of plant. You can find your hardiness zone in the United States on the USDA website. Most summer gardening plants are rated for the majority of temperate climates, but it's always nice to be sure before you plant something.

- *Plant name- common and botanical*

    - A plant tag will include the common and the botanical (scientific) name of the species and variety. Sometimes you might see a third component to the name, that is most likely a cultivar. Cultivars are just a further sub-variety of a specific species and may indicate a different color or size from its parent species.

- *Picture and description of the mature plant*

    - This is self-explanatory. It's just nice to see what your end result should look like, or what size the flowers, fruit, or vegetables are going to be.

- Planting and spacing instructions

- Knowing how to plant your seedlings and how you should space them is crucial information. You know best the dimensions of your garden, and when you're choosing seedling and starts, don't buy too many! Remember that these tiny shoots will become full-size plants over the coming months, and they will need room to grow, expand, and breathe. This is why compact varieties are best for raised beds, especially for growing vegetables. You can also grow up rather than out, and we'll go over vertical gardening before this chapter is over.

- Raised bed gardening is a great way to intermingle flowers, herbs, and vegetables in a limited space, so be sure to choose a range of different plants. We'll talk about companion planting in the next section so you can arrange all your varieties to maximize pollination, keep your soil healthy, and avoid pest infestations.

- Days to bloom/days to harvest

- Another piece of useful information you'll find on a seedling tag is how long it will take you to see your first flowers or food. This can help you make decisions about which varieties you want to plant and give you a guideline to help determine if your plants are maturing on schedule. If they're not, it's an indication that you need to do some troubleshooting. This bit of information should also include if your plants are short-blooming, long-blooming, single-harvest, or continuous harvest. This is something else to take into consideration when you choose your varieties.

- Sun requirements

- Part of being a successful gardener is giving your plants the best environment to encourage proper growth. A big part of this is making sure they're getting the recommended amount of light. Most veggies and herbs require a minimum of 6 hours of sun each day, while some ornamental species do better in partial and full shade. Knowing these requirements will help you choose varieties and put the right plant in the right place.

- Water requirements

- Most plant tags will also tell you how thirsty a plant is, which can help you decide which plants to install. Keep in mind that most gardens require approximately four inches of water each week, and that total should be slightly higher for raised beds because they usually drain faster than a native soil garden. If you've got problem areas where water pools or you're on a slight slope, you can get thirstier plants or plants that tolerate wet conditions to put in those spots. In that way, you can make the most of every inch of your limiter raised bed space.

- Indication of heirloom/ hybrid/ GMO or non-GMO

- On your seedling tags, you will find it noted whether a plant variety is an heirloom, a hybrid, or GMO/non-GMO. To choose between these options, let's take a quick look at what these terms mean. Heirloom varieties are true to species. New plants are grown from seed that is the exact genetic material as the generations of plants that came before it. Hybrid varieties are those that have been carefully crossbred by botanists to exhibit the desirable traits of two or more heirloom varieties. The characteristics of these plants can fluctuate from generation to generation because they can exhibit recessive genes.

- GMO stands for 'genetically modified organism'. These varieties haven't been bred to exhibit characteristics, they have been changed on a genetic level to eliminate bad traits and emphasize good ones. GMOs were developed for several reasons, not the least of which was to modify corn to be easier to grow in arid climates, and the popular Flavr Savr tomato, which had its ripening compounds inhibited to have a longer shelf life. GMO/non-GMO has become an ongoing argument in the past two decades, and there are proponents for both sides. You can decide on your own if you want to grow GMO varieties, but I encourage you to do some research before you purchase plants.

Once you have a basic understanding of how to read plant tags, you can begin to choose the varieties you want to grow. If you are focusing on gardening for food, you should choose compact versions of some easy-to-grow favorites to get you started.

Beans/peas- Beans and peas are legumes, which are enormously beneficial to the health of your soil. These plants return much-needed nitrogen to the soil as they grow, and they can be planted next to your heavy-feeding vegetables to prevent the stripping of nutrients. There are some fantastic bush varieties of all your favorite beans and peas available, but if you must have the climbing varieties, check out the section on vertical gardening later in this chapter.

Cucumbers/Melons- Cucumbers are a versatile plant to have in your raised bed garden because they are medium feeders, grow quickly, and the fruit can be used fresh or pickled for use later. They are a member of the cucurbit family, as are melons. You can find compact or bush varieties of cucumbers and melons, eliminating a tangle of vines and providing you with cool, refreshing produce from mid-summer through fall. One note, cucurbits require a lot of water to promote proper growth, as the ripe fruit has very high water content.

Peppers, Tomatoes, and Eggplants- You might not be aware that these three species are in the same plant family, as are the humble potato. This is the nightshade family, and they are garden staples worldwide. These plants are easy to grow and maintain, but they are heavy feeders and will strip your soil if you aren't careful. You can fertilize mid-season (more on fertilizer in a bit) and make sure they've got lots of organic material to supplement their feed. These plants should also not be planted next to each other in the garden, to alleviate an attack from nightshade-loving pests and to keep your soil healthier.

Asparagus- Adding asparagus to a raised bed garden is a choice that will teach you patience and self-control. Asparagus is loosely related to the lily family, and it is a perennial. This means that once you plant it, it will take a couple of years to establish itself, but cared for properly, it will continue to work in your garden for up to thirty years! The patience and self-control come from being able to resist eating those tender green shoots until you reach year three. Good luck!

Strawberries- A perennial fruit, strawberries are also a perennial favorite. Strawberries are actually part of the rose family, and you can find terrific compact varieties to put in your raised bed garden. You can also consider standard varieties to wind through the beds as living ground cover. Once you plant some strawberries, you can enjoy their fragrant flowers and their tasty fruit for up to ten years before they need to be replaced.

Blueberries- Compact blueberry bushes have become popular in recent years and for good reason. These plump berries are rich in nutrients and antioxidants, making them a breakfast and snacking favorite. The one catch with planting blueberries is this- they do prefer a more acidic soil than most fruits and vegetables. However, if you've got one problem corner of your garden, where you know the drainage isn't quite right, it gets a little shade, or any other tiny glitch, you can isolate a berry bush or two in that spot and give it some acid plant food to help it thrive without acidifying the rest of your soil.

Herbs- Herbs are the epitome of edible gardening, aren't they? You can plant herbs among your fruits and veggies and watch them grow and flower, repel insects, and the best part, make tasty accompaniments to all your favorite foods. It's great to be able to walk outside and snip off some fresh basil or oregano to toss into your dinner recipe. (Note: the difference between herbs and spices is that 'herbs' refer to seasonings made from the foliage of the plant, and 'spices' refer to seasoning made from any other part, such as the seeds, stems, bark, roots, etc.)

You can easily add ornamental plants to a raised bed garden to add interest, draw pollinating insects, and repel pests. There is a vast array of ornamental plants and your personal aesthetic sense may dictate what you plant, here is a list of good starter ornamentals that do well in a raised bed, and how you can interplant them as companions to your food plants.

French marigolds- Marigolds are an easy-to-maintain annual that come in a variety of colors and sizes. French marigolds are among the hardiest and there are several cultivars that do well in raised beds, such as Queen Sophia and Honeycomb. French marigolds (which actually originated in Central America) are drought-resistant, repel deer, and don't have much odor, although their color draws in bees and other pollinators.

Rudbeckia and echinacea- You may know these perennial wildflowers more commonly as black-eyed Susans and coneflowers. These are terrific pollinator plants and help prevent soil erosion with their strong roots. You can interplant them with your vegetables and herbs to create a strong ecosystem. There are many compact varieties available that will work well in raised beds.

Daffodils- Another perennial, this time a bulb. Daffodils are easy to grow, pleasant to look at, and best of all, deer-resistant. A perimeter of daffodils will bloom in the mid-to-late spring and protect your tender young seedlings from being nibbled by four-legged pests. One note about daffodils- once they are established, they will begin to spread. You can divide the bulbs every few years and transplant them elsewhere or give them away.

Calendula- This annual is technically an herb, but the flowers are beautiful and beneficial. They attract both pollinating and predatory insects, and their sticky sap traps unwanted pests like aphids and whiteflies. In warmer hardiness zones, calendula can be left in the garden over the winter as a cover crop.

Nasturtium- Also an annual herb, nasturtium has several beneficial properties. It grows low to the ground and acts as a natural mulch, its foliage is unique and provides an interesting texture, and it has a strong (not unpleasant) odor that repels pests. In addition, both the flowers and the foliage are edible.

German chamomile- If you like a delicate aesthetic, you'll love German chamomile in your garden. This self-sowing annual has tiny, sweetly-scented flowers that bring in pollinators and beneficial insects. You can leave the plants in the garden over the winter to prevent erosion and add organic material to the soil.

Most, if not all of the above fruits, vegetables, herbs, and ornamental plants will grow in warm to temperate hardiness zones. If you are in a tropical or cold zone where the growing season is much longer or shorter, you should investigate varieties of these plants that thrive in your particular zone.

Chapter 3: Easy Planting and Maintenance Techniques

Once you've chosen your plants, you'll need to get them installed into your raised beds. For planting and everyday maintenance techniques, there are some basic tools you'll need and want to have on hand. These include:

- trowel, hand rake, and weeding fork

- bulb digger

- short-handled pitchfork

- garden hoe or garden ax

- sturdy gloves

- weeding bucket

- harvest basket

- utility scissor/knife and twine

- hand pruners

- watering can or hose

- stakes or trellises for unruly plants

- tool storage shed or trunk

- a garden journal or journaling software application

Having the proper tools on hand will make any gardening task easier. If you're on a budget, check around at rummage sales and online in your local buy/sell/trade groups for bargains on second-hand tools. You can also take advantage of end-of-season clearance sales at garden centers and home improvement stores. You should use your journal from beginning to end of the season to mark down what you're planting, what products you've applied (fertilizer, compost, pesticides, etc.), how much you harvest, and other notes that will help you with dissecting and assessing the season and planning for next year.

Planting Your Seedlings

When it's time to get your plants installed in your raised beds, you want to make sure you arrange them in a way that will be the most beneficial. If you are installing any perennials, you should choose their location first, and build around them. You want to make sure you aren't placing your heavy feeders, like the nightshades, next to each other. Split them up with herbs, legumes, and your flowers, and cucurbits. This will help naturally with pest control and will discourage your soil from becoming stripped of its nutrient load.

To physically plant your seedlings, you'll want to make a hole that is deeper and wider than the pot or seedling cell that they are currently in. For small plants, you can use a bulb digger to make even holes. Be sure to use the spacing instructions on the plant tag so your seedlings have room to spread, grow, and have adequate airflow. You should give the pots or cells a squeeze to loosen up the soil and then grasp the seedling at the base of the stem and give it a good wiggle to release it from the pot.

Once the plants are free, gently massage the soil and roots to loosen them up and place them in your holes. Fill in the holes and lightly tamp down the soil around the base of the plant. When all your plants are installed, give the bed a good watering and leave the plants to settle into their new homes. Don't fiddle with them unless they look seriously askew, they will need a few days for their roots to reach out and establish themselves in their new soil. Place the plant tags in the garden to remind you which plants are which, or use a permanent marker and plastic spoons or knives to make your own plant markers.

Watering for Maximum Plant Health

One of the most important functions of a gardener is to provide water for their plants. You have taken responsibility for growing plants where none were growing before, and you must give them the necessities to thrive. I might go so far as to say that watering properly can be the make-or-break key to any garden. Cultivated plants require approximately four inches of water a week to maintain health and promote growth. You need to be able to meet those needs to be successful.

You should plan on watering your garden every day, except if you have had or anticipate heavy rain. Many people choose to water by hand with a hose or watering can, but raised bed gardens are the perfect set-up for adding drip irrigation or soaker hose system. If your hose spigot is close to your garden, these can be a relatively inexpensive way to make sure you are watering enough, and they are simple to set up.

Soaker hoses and drip irrigation work by delivering water directly to your soil and the roots of your plants. Soaker hoses are the more 'primitive' of the technology. These are hoses, capped on the far end, made of permeable material that you split and run from your spigot, through your beds, and cover with a light coating of soil. When the faucet is opening, water flows through the hose's permeable sides and into your garden. A simple timer added to your spigot can make a world of difference in having well-watered, happy plants. A drip irrigation system is similar but more targeted. Instead of being a permeable hose, there are tubes or hoses with holes that can be placed directly near the base of your plants. Both systems are available in DIY kits.

HOW IT WORKS

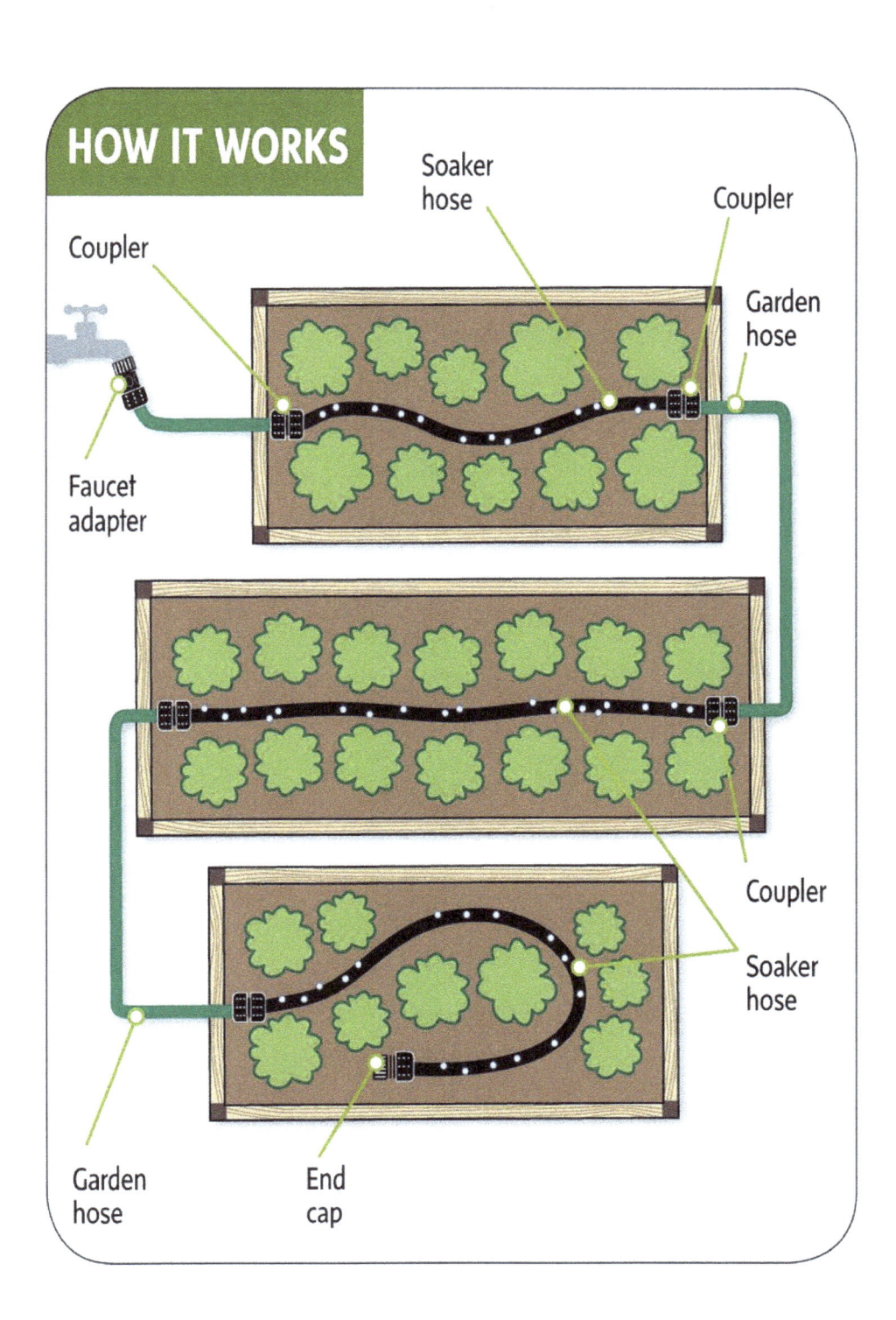

Coupler

Soaker hose

Coupler

Garden hose

Faucet adapter

Coupler

Soaker hose

Garden hose

End cap

87

- A simple DIY soaker hose system can ensure your garden is adequately watered -

If you are watering with a hose or can, be sure to always water the roots, not the leaves. You don't want your plants to get sunscald or invite pathogens to move in on wet foliage. No matter how you choose to water, a rain gauge is a good addition to any garden. They come in a ton of designs from industrial to whimsical, and they are helpful in showing gardeners of all abilities how much natural moisture your garden is receiving. You'll be surprised. Sometimes the hardest downpours don't shed as much volume as a light rain that lasts hours.

Fertilizer and Compost for Optimum Growth
Gardening in raised beds poses a unique challenge in soil health. While you may not have needed to do an initial soil test based on your supplier, you will need to do a little work to keep that soil healthy throughout your first year and beyond, because you don't want to have the expense of replacing the soil frequently. You can accomplish healthy soil year after year by liberal use of compost and discerning use of fertilizers.

You cannot ever add too much organic material to your soil, and compost is a fantastic way to achieve this. Whether you make your own or purchase bagged or bulk compost, you can add a layer to your beds at any time and it will never be too much. Compost is made up of decomposed plant matter and food scraps which is chock full of beneficial microorganisms. It helps create a thriving soil ecosystem in your garden. When you have more healthy organisms in your soil, your garden will retain more moisture, and have better soil pore space which allows for the movement of air and water to the roots and lets the roots spread for sturdier plants.

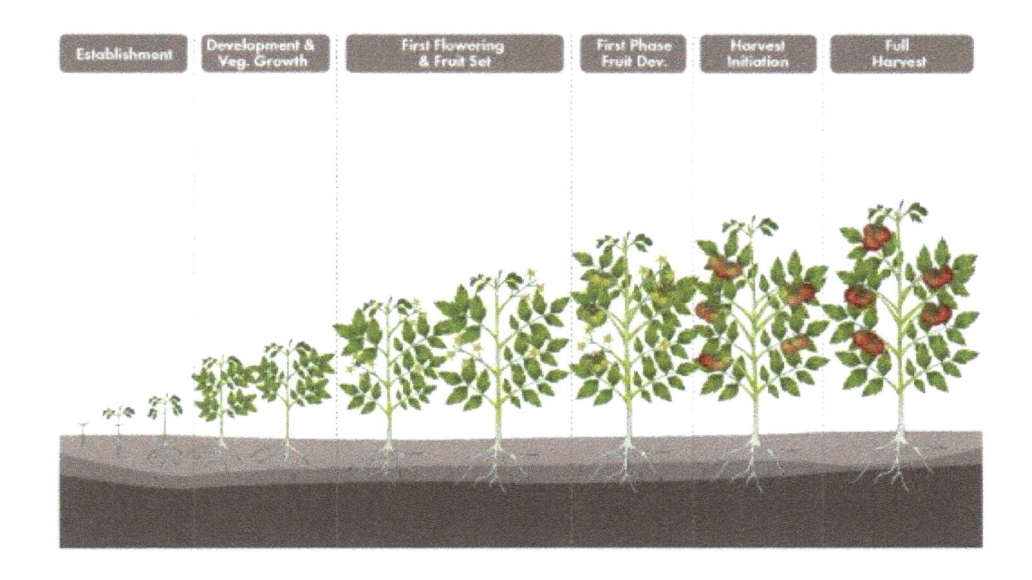

- Understanding the life cycle of your vegetables will help you know if you need fertilizer -

Fertilizer is a great tool for encouraging continued growth in heavy feeders and alleviating the drain of nutrients for your soil. However, fertilizer isn't a panacea and should be used sparingly and correctly. Too much of it will cause toxicity of nitrogen, phosphorus, and potassium (the big three nutrients), and too much phosphorus running off from your garden into the water table can contribute to larger environmental concerns.

When you use fertilizer partly through the growing system to supplement the nutrition in your garden, it is called 'side-dressing'. Be sure to follow all directions on your fertilizer package closely. If your product needs to be diluted, be sure to do so in the proper amount of water. Spray (or sprinkle) only near the roots of the plants you want to fertilize, because you don't want to encourage weed growth, either. If you have any questions about fertilizer application, you should call or email your local Cooperative Extension office or Farm Bureau. You will find someone there who can help you decipher your needs.

Dealing with Weeds, Pests, and Pathogens
One thing all gardeners have to deal with are unwanted visitors in our garden spaces. Weed, insect pests, furry critters, and pathogens are always lurking, looking for vulnerabilities to exploit. So how can you protect your plants from these invaders? Let's take a look at weeds first.

Weeds are plant matter that pops up in your garden, unplanted and uninvited. Of course, it's been said that to identify a weed, you must first know the heart of the gardener and the intent of the garden. That being said, I don't know too many people who cultivate crabgrass. Manual weeding is always an option, and some people love weeding, saying it's a great task for exercise and mental catharsis. I'm not one of those people, so here are some tips on avoiding and alleviating weeds in your raised beds:

1- Fill the space with plants you want. Don't give weeds any room to move in. Happy, healthy plants won't let too many weeds move into their territory. Use companion plants that will act as ground covers and choke out weeds before they can grow too large.

2- Mulch! If you can't cover the soil with plants, cover it with mulch. This could be grass clippings, straw, leaves, plastic or cloth weed-blocker, or traditional wood chips, although I would caution that chips should be your last choice in a raised bed garden. They don't break down quickly into the soil which could cause you issues in later seasons.

3- Reduce tilling. Weed seed likes to hide dormant in the soil and weeds begin to show up when the soil is turned, exposing them. While it is a good idea to aerate your soil with regular cultivation, it's not recommended to continuously turn over your soil exposing those weed seeds. Try to reduce your tilling to preparatory work at the beginning of the season and maybe perform a second tillage in the fall, to turn in the last shreds of spent plant material.

4- Herbicides. This should be your last resort, for a couple of reasons. First, if you are growing food, you want to be careful about spraying these chemicals on plants you intend to consume. Second, most commercially-available herbicides are not discriminatory, meaning they will kill all plant life they come in contact with, not just your weeds.

While weeds are a major concern for most gardeners, another thing to keep an eye out for is insects. Not all insects are harmful, so how can you know the difference? Well, if it's eating your plant, it's harmful, and if it's eating the insect eating the plant, it's beneficial. You want to be able to draw beneficial insects to your garden for this purpose and for pollination, and also be able to identify any harmful insects so you can take action before they can cause too much damage.

Like herbicides, heavy pesticide usage is not recommended for food gardens. There are several alternatives to using chemical pesticides in the garden, including horticultural soaps and oils, attracting beneficial insects and birds to eat the harmful pests, and in some cases, washing away the pests with a strong jet from your hose; this is a useful approach to ridding yourself of aphids and whiteflies. Try hanging a bird feeder to bring in omnivorous birds to chow down on your insect pests. You can also purchase large clutches of ladybugs to release in your garden to help with harmful insect control.

In the case of large bugs like tomato hornworms or vine-boring beetles, you can remove the pests manually and dispatch them as you see fit. You can also try sticky traps for whiteflies and larger beetle species, like Japanese beetles. For slugs and snails, use up the old beer in your refrigerator by placing the brew in shallow dishes in the garden overnight. They will help themselves to a drink, and there will be no survivors. If you're having a hard time identifying a bug, take a photo and search online. If that doesn't wield any definitive answers, email the photo to your local Cooperative Extension or Farm Bureau- they have staff and volunteers who are trained to assist with these types of questions.

The best way to stop insect issues is to be observant. Take care to look at your garden every day and note any changes like foliage and vines being eaten. Look for slug trails and insect frass (poop) on or under the leaves. Being vigilant is your best defense against pest issues. If you have tried everything and you're in need of using a pesticide, try to use organic products, especially on food plants. When all else fails and you find you have to use any pesticides, make sure you follow all label directions to the letter. You don't want to contaminate your food or the environment, and you definitely don't want to cause yourself any physical injury.

If your pest problem isn't bugs, but woodland creatures, there are a few approaches you can take to humanely deal with the problem. If small burrowers like chipmunks, rabbits, and voles are your culprits, try sonic stakes in the ground around your raised beds. These are usually solar-powered and are available at many garden centers or online. The stakes emit a high-frequency tone that isn't able to be heard by the human ear, but drives underground rodents nuts. For larger munchers, like deer, groundhogs, skunks, and raccoons, consider putting netting or fencing around your beds, and spray a pheromone-based repellant around the perimeter of your garden area.

That leads us to the third bane of the gardener's existence, and that is plant pathogens. While gardeners want to encourage all the helpful bacteria that live in the soil, they also need to avoid the harmful microorganisms that live on plants. These can be fungi, bacteria, or viruses, and the number one thing you can do to avoid infestation is to have proper airflow in your garden. Dry leaves are happy leaves. Microorganisms love to find wet, dark places to thrive. Keep your garden airy and your foliage dry to discourage colonies from forming.

Just like people, plants have an immune system that protects them from disease. If the plants are healthy and happy, they will be less prone to infection, so just by following best practices, you can help your plants stay free of disease. You should also be mindful of any 'wounds' on your plant- these can become infected just like a cut on human skin. Take care when pruning or harvesting to only remove plant material at the natural joints to avoid any opening that could invite infection.

If you find that an infection has moved in on your plants, the first thing you should do is remove the affected part of the plant or the entire plant, if necessary. Don't compost diseased plant material, dispose of it in the trash. You don't want to encourage the harmful organisms to breed in your compost pile and end up back in your soil later. The most common plant pathogens that you will see in a home garden are powdery and downy mildew, anthracnose, early and late blight, blossom end rot, and mosaic virus. Like with insect pests, you're going to want to be observant. Many times, you can save a plant or plant population by catching the disease and removing the diseased portion without losing the whole plant. This is another reason it's so important to keep plants of the same family separate in the garden. You don't want powdery mildew to affect all your cucurbits or blight hitting all your nightshades.

If you find that you must use more aggressive methods to get rid of pathogens, be smart about your approach. You can get a wide variety of organic fungicides, antivirals, and antibiotics to treat your plants, but always be sure to follow all instructions to the letter and use personal protective equipment if it's indicated. One note about using these products in a home garden: most vegetable plants are annuals and only have one life cycle. If a plant is so diseased that you think commercial remedies are necessary, it's probably just easier to throw away the plant and try again next year. It's better to expend your time and money saving valuable perennials if you must use disease-killing products.

SIX COMMON VEGETABLE GARDEN DISEASES

LATE BLIGHT
Light green or brown spots on leaves. Fuzzy growth and discoloration on stems and leaves.

PLANTS AFFECTED: tomatoes and potatoes.

EARLY BLIGHT
Brown "bullseye" surrounded by a yellow ring on leaves and lesions on stem.

PLANTS AFFECTED: tomatoes and potatoes.

POWDERY MILDEW
Light colored "powder" on plant leave, flowers, and fruit.

PLANTS AFFECTED: beans, grapes, melons, cucumbers, and squash.

DOWNY MILDEW
Yellow or brown patches on leaves and they may wither or curl.

PLANTS AFFECTED: melons, squash, pumpkins, cucumbers, and broccoli.

BACTERIAL WILTS
No noticable discoloration but plant constantly looks wilted despite care and watering.

PLANTS AFFECTED: cucumber, gourds, pumpkins squash, and muskmelon.

RUST
Yellow, brown, and orange spots on leaves that turn into bumps. When leaves are rubbed or crushed the plant creates a reddish dust.

PLANTS AFFECTED: beans, leeks, onion, and corn.

- Being vigilant will help you identify issues and handle them before they explode -

I can't impress upon you enough how much vigilance and observation are key to identifying and eliminating pests and pathogens in your raised beds. It's also important to follow best gardening practices and do things that will invite helpful organisms, like good bacteria in the soil and beneficial pollinators and predator insects to control the pest population. This practice is known as IPM, or integrated pest management, and it helps you build not just a thriving garden, but a thriving ecosystem.

Chapter 4: Maximize Space with Vertical Growing

If you have your heart set on growing things that don't come in favorable compact varieties, or if you just want to maximize the space in your raised beds, you can utilize vertical growing to suit your needs. Vertical growing is precisely what it sounds like, and it can be easy and fun!

You can use almost anything your heart desires as a trellis or support, and it's entirely up to budget and taste. One great idea I've seen over the years is using a repurposed children's soccer practice net- you know, those square string grids on a metal frame. It was set in the center of a raised bed with peas and beans climbing it. You can make something similar with just a few lengths of lumber and some sturdy twine. You can also use the handles from old tools (broken rakes, anyone?), tomato cages used both narrow and wide side down, and bits of lumber like one-inch furring strips. Be creative!

Another vertical idea is to place tall shepherd's hooks in the corner of your garden to hang pots on to drape the plants downward, or to build a four-corner frame, like a four-poster canopy bed to fit the dimensions of your bed. You can grow things up all the legs and along the beams to get a ton of growing space that you didn't have before. So, what can you grow vertically? Anything vining variety! This means you could free up valuable horizontal space by growing your cucumbers, melons, strawberries, peas, and beans, as well as vining ornamentals, on trellises and supports. The only requirement is using materials sturdy enough to hold those plants once they begin to get heavy with fruit and flowers. And of course, a step stool, if you need one. I once had to harvest cucumbers out of my dogwood after a particularly ambitious vine surpassed its trellis, the stockade fence beyond it, and began to climb the tree!

Chapter 5: Harvesting and Overwintering

Growing and maintaining a garden is great, but getting to harvest and enjoy the things you've grown is even better. When you grow your own food and flowers, you can always have a home full of fresh beautiful things. Because how you harvest is important as when you harvest, let's go over some basics for both food plants and ornamentals so you make the most out of your garden while it's at its peak.

When you cut flowers for bouquets or arrangements, it is always best to cut the stem at a length longer than you need- you can always trim the excess later. Try to cut at a joint or at the base of the stem, and always use a pair of sharp hand pruners or utility scissors to make a clean cut. You should also make sure your cuts are on the diagonal. This lets the plant heal faster and gives the cut end a larger surface to take up water from your vase or floral foam. Have fun arranging your cut flowers! It's such an easy way to brighten a room or brighten someone else's day! A birthday bouquet is that much more special when the flowers are homegrown.

For harvesting your food crops, you should refer back to the information on your plant tags, which will let you know approximately when you should expect to have your first pickable crops. Some varieties will continue to fruit for the entire rest of the season, as long as you harvest, they will keep producing. Some are more of a one-and-done crop and will give you their harvest and then begin to fade. Once your garden begins to get close to the mark for beginning to harvest, then you should keep an eye out every day for ripe fruits and vegetables. If you're unsure, it's always best to pick a little early than a little late.

When you harvest your fruits and veggies, it's all about clean removal to avoid opening a 'wound' on the plant. You should always pluck things like beans and peas from the stem joint above the pods, not right at the pod itself. Cut cucurbits from their stems at the closest joints, not directly at the top of the fruit. Tomatoes will loosen and 'fall' into your hand when they are ripe. If you do damage any of your fruit or veggies when you harvest it, be sure to eat or cook and refrigerate the damaged goods that same day.

If you're not going to be eating your fruit and vegetables fresh, you should make sure to can, freeze, dry, or otherwise preserve it as soon as you can to make sure you've not lost any nutrient content of the produce. When you freeze veggies like beans and peas, you should blanch them first to seal in the antioxidants. Don't boil them; this will leech the nutrients out of the vegetables and make them mushy when you go to thaw and reheat them. Blanching retains the nutritional value and makes the veggies useful for a variety of cooking methods, including things like soups and stir-fries.

After you've harvested and the garden is spent, it's time to prepare for winter. You want to make sure that you do everything possible to have healthy soil for the next growing season, and there are a few steps you'll need to take to put the 'bed' in raised bed. The first thing you should do is remove the remains of all annual food plants. You don't want unwanted seeds falling into the soil, leading to surprise plants next year.

Next, you should remove any annual ornamental and herb matter. Remember, you can compost any spent plant material that is free of disease. After all your annual material is removed, you'll want to cut your perennial flowers down to no more than two inches above ground level. Prune any perennial shrubs, ornamental or otherwise, and cover them with burlap, secured with twine. Trim any dead matter off of perennial ornamental vines. Once you've attended to all your plant material, it's time to take care of your soil.

Because your soil needs to last you for several seasons without being replaced, you want to add as much organic material as you can when you put your garden to bed for the off-season. Give the soil a hearty layer of compost, turning it in with a pitchfork. Stab the heck out of the soil while you're doing this- it will aerate the soil and break up any clumps that may have formed around the roots of your plants. Once you're done composting and aerating, you can either plant a cover crop, like winter rye, hairy vetch, or red clover, or you can lay a thick layer of straw (not hay!) in your beds. Now, let the beds sleep. If you have a dry winter, you can water the beds occasionally to encourage decomposition.

In the spring, you're going to want to turn all that fabulous organic material right into the soil and prep it in the same way you did when you first built the beds. Make sure that you practice good crop rotation techniques when you plant your garden again. This means moving everything one spot to the right (or clockwise) so that your heavy feeders are not in the same spot they were last year, and your legumes have a chance to replenish the soil in a new spot. Crop rotation will give your garden the opportunity to heal and regulate itself and give your plants the healthiest soil in which to grow and thrive.

One last note about raised bed gardening is that you've got to make sure to maintain the beds themselves. If you've built your beds from wood, watch for rotting or splitting, and tend to any damage immediately. You should also keep an eye on any garden structures made from masonry block, for cracks and crumbles. You don't want to have any mishaps, so it's always best to nip any structural damage in the bud. Be proactive about maintaining your beds and they will treat you to years of gardening happiness. Have fun and be creative with your design and planting, and you'll have great success at being a raised bed gardener.

PART IV

Urban Gardening

If you live in an urban setting, you might think that you can't be a gardener, but nothing could be further from the truth. As long as you can find some sunshine, you can be a gardener, whether you've got a balcony, a courtyard, or an entire empty lot to work with. Urban gardening can be an exercise in creativity and will reward you with food and flowers you can enjoy with family and friends, no matter how much or how little space you have. Let's start by taking a look at the special design considerations for creating a beautiful garden oasis in a concrete landscape.

Chapter 1: Planning and Building an Urban Garden

Living in a city means there often isn't a lot of open soil available for gardening unless of course, you go start tilling up a plot in a public park. Since that's likely illegal or at the very least, frowned upon, the first thing you need to do when designing an urban garden is to define your space.

If you've got a patio or balcony, then you will be able to do your gardening in containers or small growing tables. For larger spaces like courtyards, you may be able to construct small raised beds. Many residential stand-alone homes in urban areas have retaining walls built into their driveway areas, so you may be able to take advantage of these structures and carve out some terraced gardening space along these walls. Think about all the possible places that have gardening potential and observe them to see how much sun they get each day.

Another consideration for urban garden design is what's known as the heat island effect. Cities are made up of a lot of impervious surfaces like glass, concrete, asphalt, and brick. These materials collect heat during the day and release it at night, which is one of the reasons that large cities are warmer than their immediately surrounding suburbs and small towns. When you garden in a city that creates a heat island, you have to adjust to the temperatures and growing conditions that the effect creates. Because of these elevated temperatures, urban gardens require a lot of watering, so you'll need to make sure you've got a handy water source.

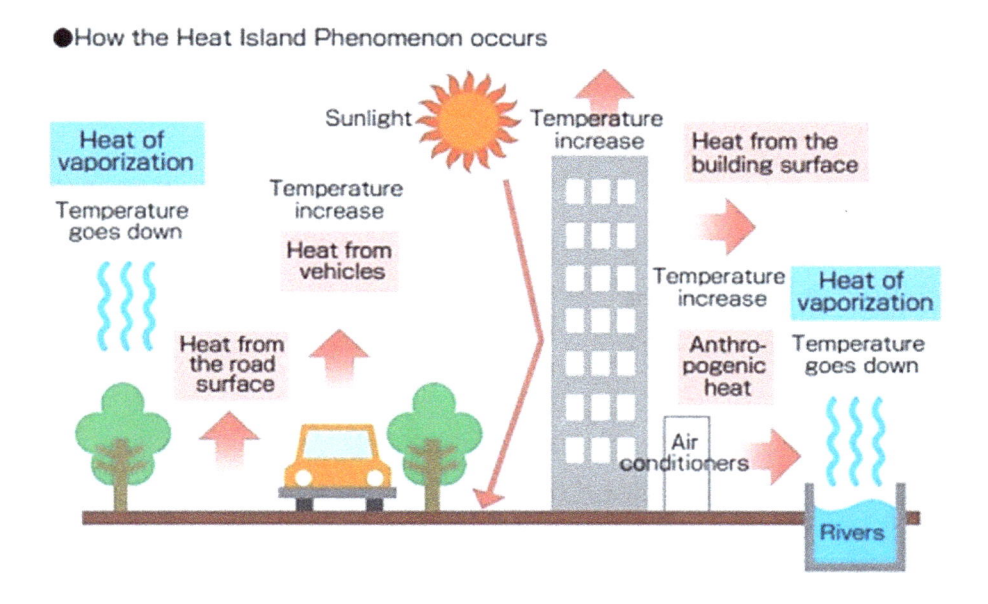

●How the Heat Island Phenomenon occurs

- How the urban heat island effect works -

Once you have chosen a location for your gardening endeavors, it's time to have fun designing! If you're going to be using containers, you have a lot of flexibility about what you can do. Depending on what you want to grow, you can use as few or as many containers as you'd like, given that your smallest containers are large enough to support one plant (generally eight to ten inches in diameter). You can use larger containers to group several plants together, as well.

For container plantings, height is another factor that adds dimension and interest to an urban garden. You can use small tables or homemade wooden platforms to set your pots at varying heights and make use of hanging baskets to add a top layer of depth and maximize space. You can also find commercially-made stacking and step-style planters that offer a lot of growing surface and can fit in a compact space.

If you have room to build small raised beds or install growing tables or troughs, you also need to consider the heat island effect. Placing raised beds on surfaces like asphalt and concrete means that the soil temperature in the bed will be warmer than the soil would be in a traditional native soil garden. This isn't necessarily a bad thing- it can extend your growing season and allow for succession planting, but it does mean that your garden will also be thirstier. Raised beds that are set on impervious services are most effective when they are at least two feet deep, so consider that when you are thinking about designs.

Whether you are using containers or small raised beds, you should also be mindful about placing them directly next to a building, because the building will also radiate collected heat during the night. In colder climates, this is actually a useful effect of the urban heat island, keeping tender seedlings alive during chilly evenings. In warmer climates, it can have a wilting effect on plants, who need the cooler temperatures at night to rest from extreme heat during the day.

It's recommended that no matter what type of containers, tables, or beds you decide to install, you should make them non-permanent structures, and if you are renting your home in the city, you should check with your landlord or building officials before placing any pots or raised beds. This is protective of both you and the property owner or manager. You don't want to put in a lot of effort to install things that you may be told to take down, you don't want to violate any safety or building codes, and you will also want to make sure that it's clear whether or not you intend to leave behind or take with you any gardening structures should you decide to move when your lease is up. Get everything you can in writing so there's no confusion later.

You should measure your space and sketch out your designs with pencil and paper, keeping in mind that you'll need to conserve some space around your containers or beds to have room to work. It's a lovely thought to want to cram as many plants as possible into a small area, but you want to be able to water, weed, and harvest without needing to be a contortionist. You also want to make sure your garden has good airflow. Draw a handful of designs and see which you like best, and then it's time to choose your materials.

This part of urban gardening is really fun because it can be like a scavenger hunt, both in real life and online. Try to find reclaimed materials around your city- look at rummage sales, at online buy/sell/trade and freecycle boards, and connect with the local waste authority to see if you can get discarded construction materials. You can get pots and containers that fit your tastes from quiet to quirky, and find lumber and masonry block to build your raised beds if you've got the room. If you want all your pots to match, you can also save cash by purchasing in bulk. Pre-fabricated bed kits and table kits are also an option for those who aren't inclined to DIY construction projects.

Once you've gathered your materials, you can get to assembling your garden. If you're building raised beds, make sure you've got all your tools at the ready before you begin. If you're using pre-fab kits, check to see that you've got all the parts and necessary tools on hand before you begin assembling. You don't want to get halfway through a project and realize you're short some hardware or don't have the right screwdriver. If you're cutting materials, always measure twice and cut once, as they say, to avoid goofs and waste. If you are assembling pre-fab kits, save the paperwork in case you ever need to replace any hardware, and if you've constructed your own beds, save your plans and sketches so you remember what materials you've used if you need to do any replacement or maintenance down the road.

With your raised beds built or your containers gathered, the next thing you will need is some soil. It may be difficult to get bulk soil delivered in the city, but it's worth looking into the option just in case. If you cannot, you're going to have to go with bagged soil. For containers, you can go with straight potting soil, and for raised beds, you'll want to use a 50/50 mix of potting soil and topsoil. For reference, potting soil is lighter than topsoil and contains a mineral mixture to help your plants get the necessary nutrients. Topsoil is, well, dirt, although I don't like to use that word. It's mineral and nutrient content is often inconsistent, but for filling raised beds, it will give you more bang for your buck than the expense of potting soil alone. You can always amend the soil to make it more nutrient-rich (more on that in a bit!)

The reason you want to get a good potting mix for containers is twofold. Potting soil, being lighter than topsoil, will mean that your pots won't get so heavy you shouldn't be able to move them around if necessary, although large pots can always be set on wheeled dollies if you want to be able to roll them around. The other reason is that potting soil contains nutrients to get your plants off to a good start and through much of the growing season, although heavy feeding plants may need a bit of a fertilizer nudge to avoid a mid-season slump. Look for potting soils that have sphagnum or peat moss, and vermiculite and/or perlite. Some mixes also have fertilizer built right into the product, so if you get these, you're not going to want to use additional fertilizer later, or you may risk plant toxicity.

Okay! Are you all set up? Got your garden planned, assembled, and chosen your soil? Wonderful! Now it's time to talk about plant selection and choosing the best varieties to thrive in an urban garden. You've probably got a good idea of what you would like to grow, but let's take a look at choosing seedlings and talk about some of the plants that do well in containers and small beds.

Chapter 2: Choosing Plants for Food and Fun

Many urban gardens are small, and that means that you've got to be creative about the use of space. In this section, we're going to examine what you need to know about reading plant tags to choose the best seedlings, talk about companion planting for maximizing room in your beds and containers, and go over some of the food and flower varieties that do well in limited space.

When you're planting an urban garden, it's important to look closely at plant descriptions on the tags so you know you're getting varieties that will thrive in smaller spaces. You should look for any tags that indicate 'good for containers' or 'container-friendly'. Also, look for indicators like 'compact' or 'low-growing.' Plant tags are a wealth of information and will provide you with the following data:

Hardiness zone/growing zone: This lets you know if a plant is suitable to be grown in your locale. You can find your zone on the USDA website in the United States, or your country's department of agriculture site in other locations. You can also call that department's toll-free number for assistance. This is useful if you are browsing plants in a catalog or online. Most garden centers and retail nurseries will only sell seedlings that are appropriate for your area.

Plant name, common and botanical- This will indicate the name of the plant, the variety, and the cultivar (which is a fancy way of indicating a particular color or another identifying trait). This will help you choose a variety that is appealing to you and will fit into your garden scheme.

Photos of the grown plant/produce- A picture of what you can expect when the plant is fully grown and is producing flowers and fruit is useful in knowing if your plant is growing properly. It also helps you decide if a plant is aesthetically pleasing to you.

Spacing requirements/plant dimensions- This is vital information for planting an urban garden with limited space. If you're going to be planting in raised beds, you know your dimensions and can use this data to decide how many seedlings you can fit into your garden. When you know how far apart the plants need to be and how large they will get, it's easy to avoid overbuying and overcrowding. For containers, it's useful information because it can help you decide how many plants to put in a pot, based on dimensions. Plants should be in pots of at least 8" by themselves, but you can certainly plant more seedlings together in larger containers!

Sun and water requirements- Most vegetable and herb varieties require a minimum of six to eight hours of sun per day, while ornamentals like flowers will vary. You can use this to your advantage by choosing varieties that may thrive in that one corner the sun doesn't quite reach all day or that shady spot you just couldn't avoid having as part of your garden. The watering requirements are important, too, because you don't want to over- or under-water an urban garden. We'll go over some great watering techniques in the maintenance section of this chapter.

Days to bloom -or- harvest- This tidbit lets you know how long you'll have to wait to pick your first flowers or vegetables. It also gives you a good baseline indicator that your plants are growing on track with the norm. While you may see some action before or after the length of time indicated, this is an average for the species and variety. This may also tell you if the plant is a continual harvester, how long it blooms or produces fruit, and when to know that the plant is spent, in the case of annuals.

Heirloom/hybrid/GMO or non-GMO- While this may not play much into your decision, you should be aware of the differences between these three types of plants. An heirloom variety is one that was bred from the seeds of identical plants. This means each generation of the plant has one common DNA sequence, and planting a seed from an heirloom will result in the same plant growing. A hybrid plant is one that has been cultivated from two or more strains of DNA, and the results are a plant that exhibits the best traits of the parent plants. This is often done to increase yield and disease resistance, or to produce a new color. If you save a seed from a hybrid and plant it, you may get the same plant, or you may get one that exhibits traits from its parent varieties. GMO stands for 'genetically modified organism', and you can make your own decisions about whether you want GMO or non-GMO plants, but here is some background. The original GMOs were created for commercial agriculture to increase yield, lengthen the shelf life of ripe produce, and produce drought-resistant crops that would excel in arid regions. These varieties have had their DNA altered to be able to do so.

With all this information under your belt, you can be a pro at reading and interpreting plant tags and choose the best plants to fit your budget, your space considerations, and your individual tastes. When you're planning out your plant locations, you should think heavily about companion planting, meaning that where you put your plants is just as important as which plants you choose. You can intermingle your flowers, herbs, fruits, and vegetables to be advantageous to you and to each other.

COMPANION PLANTING

Plant these vegetables together to make the most use of space and deter pests.

BEETROOT: Brussel Sprouts, Broccoli, Onions, Cabbages, Swiss Chard

CABBAGE: Brussel Sprouts, Tomatoes, Kale, Broccoli, Swiss Chard, Spinach

CARROTS: Cabbages, Lettuce, Leek, Onions, Radishes, Peas

POTATOES: Corn, Cabbages, Beans, Peas, Squash

TOMATOES: Onions, Cabbages, Carrots

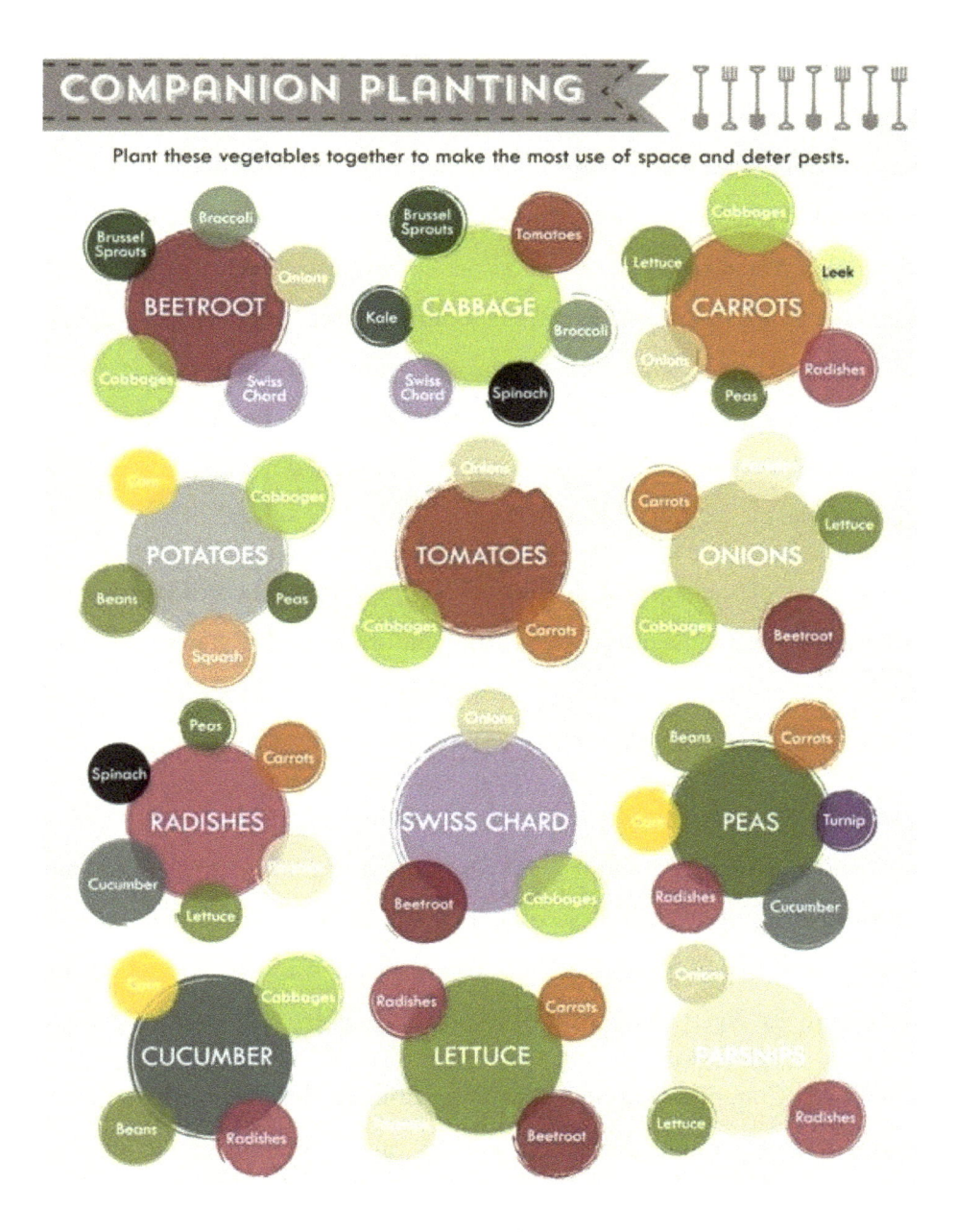

ONIONS: Carrots, Lettuce, Cabbages, Beetroot

RADISHES: Peas, Carrots, Spinach, Cucumber, Lettuce

SWISS CHARD: Onions, Beetroot, Cabbages

PEAS: Beans, Carrots, Corn, Turnip, Radishes, Cucumber

CUCUMBER: Corn, Cabbages, Beans, Radishes

LETTUCE: Radishes, Carrots, Beetroot

PARSNIPS: Onions, Lettuce, Radishes

- A handy companion planting chart to help you maximize space -

Companion planting in an urban garden is a good way to make the most of limited growing space, repel insects, and use low-growing plants as

113

living mulch for larger varieties. In containers, it's a fun idea to add the herbs you would use in cooking or preparing the vegetables around the base of the veggie plant. A good example of this might be to put oregano and basil in a large pot with a tomato plant or dill with your cucumbers. It also helps you keep some modicum of separation between your plants of the same family or species, which can 'confuse' insect pests who may attack all similar plants. Companion planting can also be aesthetic, as in when you're creating containers of flowering ornamentals. There, you want to mix varieties of differing heights, colors, and foliage textures for striking visual appeal.

Let's take a look at some of the plants you might want to consider for your urban gardening adventure. These tried-and-true garden staples come in compact varieties that will do well in any small space and provide you food and beauty all season long.

Tomatoes- These tasty members of the nightshade family are a global favorite, and many container varieties thrive in urban gardens. You can find tomatoes of all sizes and colors that will grow well in containers and small raised beds, given that they have enough space to spread strong roots, and are given plenty of healthy soil and water.

Cucumbers- Cucumbers are another favorite in gardens everywhere. They make perfect summer snacks, go well in salads, and of course, can be pickled. This hardy member of the cucurbit family can be found in bush varieties, or you can choose vining varieties if you're interested in doing some vertical gardening (check out the upcoming segment on that!)

Beans/peas- These versatile members of the legume family are a great addition to any garden, but the wealth of compact varieties are perfect for small urban endeavors. Legumes even have a superpower! They add nitrogen, a much-needed macronutrient, back into the soil, which means planting them next to heavy feeders like tomatoes in a raised bed benefits both the plants and the soil. As a bonus, legumes have delicate flowers that add beauty and a pleasant aroma to your garden, and beans and peas can be eaten fresh, blanched and frozen for later consumption, and dried for use in soups and stews.

Peppers- Another favorite! Peppers are also members of the nightshade family, so you shouldn't plant them directly next to your tomatoes. The great thing about peppers is that they come in a style and flavor for every palate from the mild and sweet bell pepper, up to the hottest of chilies and reapers. The heat in a pepper is determined by its capsaicin content, which is found concentrated in the seeds, not in the flesh. A pepper's color is also not determined by the variety, but by how long it's left on the vine to ripen. You can pick them when they are green, but they will eventually turn to yellow, to orange, to red, if you let them.

Strawberries- These sweet perennial favorites are an edible member of the rose family, and they are a fantastic choice for urban gardening because of their versatility in growing conditions. You can put strawberry vines in pots, in hanging baskets, and in raised beds to act as a ground cover/living mulch for other plants. Even once they stop producing fruit for the season, the shiny foliage is interesting and attractive.

Herbs- Growing herbs in an urban garden is easy and fun! You can plant them nearly anywhere as standalone or companion plants, and they will thrive, although most herbs tend to be thirsty. The other great thing about herbs is that most are perennial, and if you grow them in small pots, you can bring them inside during the off-season and enjoy them year-round. Species like mints and basils also draw in beneficial insects and repel harmful ones, and sages and bergamots draw in pollinators, which are important for the health of your garden.

Begonias- With their shiny leaves and delicate, bright flowers, begonias are a terrific choice of annuals for your urban garden. These plants are hardy, fairly critter-resistant, and are always priced reasonably for any budget. They come in several colors, do not grow very tall, and work well in containers and raised beds for a pop of color and foliage texture.

Impatiens- These flowering annuals provide a lot of versatility in any garden. They are almost like goldfish- they will grow to the space you give them. Impatiens have unique red tints along their stems and will produce their signature four-petal flowers continuously in a wide variety of colors from mid- to late spring through early fall in a temperate climate. They can be planted in hanging baskets, as well, and they plant well in a mixture of other annuals.

Coleus- This foliage plant looks like something out of prehistory and comes in so many colors and patterns, it would be impossible to name them all. Coleus grows straight to a medium height, doesn't take up much horizontal space, is a fun accompaniment to flowering annuals, and will flower and go to seed late in the season, so if you don't want it self-seeding itself, you should remove it before it bolts.

Sweet potato vine- The flowering variety of sweet potato vine doesn't produce any food, but it is a lovely ornamental that can be used as ground cover, or in containers and hanging baskets. The dark, almost-black foliage and its lavender bell-shaped flowers are a nice contrast to break up the endless shades of green in any garden. They are hardy, drought-resistant, and tend to be left alone by most critters, save some deer.

Celosia- Celosia is a bright addition to an urban garden, and give containers and beds something with a fun texture to play off of. The rocket varieties of celosia (sometimes called cock's comb) are best suited for small spaces, and they grow in cone-shaped, feathery bursts of reds, oranges, and yellows. They are hardy and will bloom for the majority of a growing season, often lasting until after the first frost.

Snapdragons- Like rocket celosia, snapdragons add vertical texture and interest in container gardens. They come in a variety of bright and pastel colors for any taste, and for a little bit of an edge, leave them once the flowers have fallen- the foliage pattern left behind looks like little skulls. So funky!

Bulbs- It's entirely possible to add perennial bulbs to an urban garden, and they make a great choice along the outer edges of raised beds and to bring a pop of color to your containers in the early spring. You can choose bulbs like tulips, daffodils, and irises in raised beds, and smaller, more shallow-rooted bulbs like crocuses or grape hyacinth in containers.

Other annuals you might want to consider for your garden might be lobelia, marigolds, nasturtiums, petunias, pansies, and artemisia. These all can be planted in containers and do well in hot conditions. Once you've chosen all your varieties, you can actually get down to the business of gardening. Installing your seedlings is your next step, so let's go over some of the tools and techniques you'll need to get your garden from planned to planted!

Chapter 3: Planting and Maintenance Methods for Urban Settings

The nice thing about planting an urban garden is that you don't need too many bulky or long-handled tools. A good set of hand tools, like a trowel, hand rake, hand hoe or garden ax, bulb digger, and a weeding fork will set you up well for all your cultivation tasks. You'll also want a sturdy pair of gardening gloves, some utility scissors and twine, a bucket for weeding, and a basket of some sort for harvesting. You'll also need a hose or watering can, of course.

A good canvas bag should be all you need to tuck your tools away, and it's always recommended that you keep a garden journal. You can make notes about what's working and what isn't, which varieties you like or don't like, and jot down the dates of your planting, harvesting, and if you make any fertilizer or other applications to your soil. You should also have some stakes or cages on hand if you're growing any tomatoes or vining varieties of other species.

If you're planting in raised beds and haven't used enriched potting soil in your 50/50 mix, you'll want to do a little preparation before planting. You should get your hands on some straw or old newspaper (more likely in the city). Shred the newspaper and pile it into the beds, and then wet it down so it doesn't blow away. Do this every day for a few days until you can turn the decomposing newsprint into the soil. It will add much-needed organic material to your raised beds.

For planting exclusively in containers, you won't need to do any additional soil preparation since you've filled them with potting mix. When it's time to install your seedlings, you'll need your trowel and gloves. To make your life easier, you can lay out all your seedlings in their places before you start digging. When you remove seedlings from their pots or seedling cells, do so gently. Give the pot a squeeze to loosen the soil and then get a hold of the seedling at the base of the stem. A few good wiggles and you'll have it free. Holes should be dug a little wider and deeper than the pot the seedling was in.

Once your seedlings are free, give the roots a quick massage to break up any compaction. You don't need to remove all the original soil from the roots, as this is the soil they are used to, and it will ease their transition into your beds and pots. Place each seedling upright in its new home, and fill the hole around them, pressing lightly around the stem to make sure they stay standing straight. Once all your plants are where they need to be, water them generously and let them settle in.

"Water generously" is a phrase commonly associated with urban gardening. The nature of the urban heat island phenomenon, combined with the fact that containers and raised beds drain faster and retain less moisture than native soil means that you'll be doing a lot of watering. If you have a hose, that's great! You'll be done in no time. You can even consider putting drip irrigation or soaker hoses in a raised bed, to do your watering for you. These systems are fairly inexpensive and require running the tubing through the garden bed and setting your hose spigot on a timer. Sprinklers are not recommended for urban gardening, as they are not targeted enough and tend to wet the leaves, rather than the roots of the plants. This can lead to sunscald and pathogen issues. Best to use a hose.

For small balcony gardens, you can get by with a watering can, or you can put self-waterers in your containers. You can buy the glass bulbs, or you can make some from old beverage bottles; long-neck wine and spirit bottles work well and are durable, but plastic water and soda bottles can be used, too. The physics of this is simple- you fill the vessel and turn it upside-down into the soil. Water will only flow out of the bottle when the soil is dry enough for there to be pore space for the water. Once the soil is saturated, no more water can escape until the soil begins to dry out again. This technique can work in raised beds, too, if you are willing to use larger bottles and place them throughout the beds.

If you're not putting in a self-watering system of any kind, you should plan on watering every day, at least once a day. You should always water first thing in the morning, and if it's a particularly hot day, again in the evening once the sun has moved low in the sky. Urban gardens are thirsty gardens, and it's imperative that you keep the soil as evenly moist as you can by being a vigilant waterer. Of course, if you get a heavy rainstorm, you can skip that morning, but the point is more that staying proactive is better than having wilted plants that need to catch up. And again, water the roots, not the leaves to avoid moisture issues on your foliage.

Once you've established a good watering regimen, you're well on your way to garden success. The next things to worry about are the gardener's nemeses- weeds, pests, and pathogens. Heave a big sigh, it's okay. Let's talk about these problems in that order and get you set to tackle any issues that might head your way. It's important to know what to look for, how to handle concerns promptly, and how to avoid them in the future to be a better, more experienced gardener.

Weeds, thankfully, are not usually a terrible concern in a well-tended urban garden. Because you've imported your soil, you shouldn't have to worry about the weed seeds that lurk, dormant, in traditional native soil gardens. If you take care not to let weeds get ahead of you, manual weeding a couple of times each week should rid you of any pesky intruders. Another nice thing about weeds in containers and small raised beds is that they rarely have time to root very deeply and they are easily seen and plucked. You can also lay mulch in raised beds, if you're so inclined, by laying a barrier of plastic sheeting or old newspaper around the base of your plants.

That brings us to insect pests, and that can be an issue in urban gardening. This is because plants where they aren't expected, like the middle of a city, can draw in bugs that might normally not be there. Common garden pests include aphids and whiteflies, and these tiny invaders will eat almost anything green and growing. If you see them move in, especially on your vegetables, hose them off. A good blast of water is usually enough to shake them loose and send them to a watery grave. Alternatively, you can try a horticultural soap or oil.

What's Eating My Plant?
How to Recognize Common Pests by the Leaf Damage They Cause

Damage	Pest
Deformed leaves, sucking damage	Aphids
Discolored leaves, sucking damage	Thrips and mites
Chewed or skeletonized leaves	Beetles, caterpillars, and sawflies
Leaf galls (abnormal plant growths)	Cynipid wasps, certain aphids, psyllids, and mites
Leaf mines (white patterns on leaves)	Beetle, fly, or moth larvae
Folded leaves	Caterpillars, tree crickets, and spiders
Rolled leaves	Certain mites or some caterpillars
Chewed leaves, slime trails	Slugs and snails

- Knowing what's eating your pests can help you solve the problem quickly!
-

You can manually remove larger pests, like tomato hornworms and slugs (wear gloves!), or put out beer traps in the evenings. Other common pests to look for are vine borers and beetles. Sticky traps work well for these types of pests. Another thing to consider is hanging up a bird feeder or two. These will attract omnivorous birds who will help you out by eating the things that are eating your garden. You can also consider purchasing a clutch of ladybugs, who will help rid your plants of other, unwanted, insects.

Using biologic controls for insect pests is part of a horticultural practice known as IPM, or integrated pest management. It helps farmers and gardeners reduce the need to use harsh chemical insecticides and pesticides and maintain a healthier ecosystem. One last thing to consider is planting what's called a 'trap crop'. This is something that's particularly inviting to insect pests; for instance, flowering nettles are incredibly attractive to aphids. You could plant some in a pot off to the side and let the bugs feast on that, saving your other plants from harm. One of the best things you can do for your garden, though, is to be observant. If you take a close look at your plants every day, you are much more likely to see a problem at its very beginning stages and be able to nip it in the bud (pun intended!)

Pathogens are the last gardening concern we'll address in this segment, and because bacteria, viruses, and fungi are so widely varied, it would be difficult to talk about them all, so we'll go over some basics. One of the best ways to keep disease out of your garden is to keep being a proactive gardener. Plants have an immune system that is kept strong by being healthy, so when you follow best gardening practices, you are giving them the best chance to be disease-resistant. Keeping leaves dry, not over- or underwatering, and providing proper nutrition are great ways to help your garden avoid infection from pathogens.

Some common pathogens that can plague vegetables are early and late blight, blossom end rot, downy and powdery mildew, mosaic virus, and anthracnose. Some of these can also affect ornamentals, which can also be victims of leaf spots and vascular wilts. Again, best gardening practices should help you avoid any major issues, but some diseases have a way of sneaking in no matter how hard you try. If you see that your plants are in distress, first determine if they need water or a boost of fertilizer. Some heavy feeders will begin to droop mid-season, and a proper dose of a good organic fertilizer will help them perk back up. Be sure to follow dosing instructions to the letter.

While it's rarely seen in urban gardens, due to the soil used, nutrient toxicity can also present similarly to disease, with yellowed, curled leaves or 'rusty' spots. This likely isn't the case for you, but you can certainly take photos of your plant's unhappy spots and search online. No matter what, if you've got leaves and stems that are looking out of sorts, the best thing to do is to cut away all the diseased material and dispose of it in the trash, away from the garden. Clean your pruners or scissors with rubbing alcohol and hot, soapy water when you're done to avoid infecting another plant. If you have to, remove the entire affected plant. Keep a close eye on the plants around it, and if you must resort to using antibiotics or fungicides, be very mindful of applying them using proper doses and wearing gloves.

Take care not to splash any unaffected plants, you don't want to cause any collateral damage.

Now that you've got some solid basic knowledge of planting, watering, and maintaining your garden, let's talk about the future. In the next few segments, we'll cover how you can truly maximize your garden space and increase its productivity, how you can keep your garden and soil resting and happy in the off-season, and how to take your small concept from your patio to your community to bring the joy of fresh food and flowers to your neighborhood.

Chapter 4: Using Small Spaces to Your Advantage

Just because your gardening space is small, doesn't mean your gardening yield needs to be. We already talked about companion planting to maximize your horizontal space, so now let's go a little more in-depth about vertical gardening. We'll also take a look at succession planting, which can help you get more gardening out of less space by planting a second crop during a single growing season.

If you've incorporated hanging baskets into your gardening design, then you've already started vertical gardening. By utilizing another dimension of your gardening space, you can increase the yield of your garden. Trellises, stakes, and tomato cages are useful tools for training your plants to go up, not out, and they also keep delicate leaves from laying on damp soil, which encourages pathogens. Be creative about what you use for your vertical gardening. Anything can be a stake if you put your mind to it.

Picking the Right Method

Choose between these three different succession planting methods to help your garden grow across the entire season:

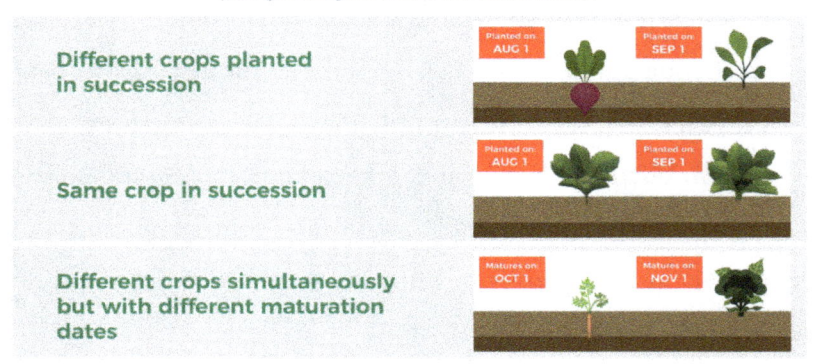

Different crops planted in succession

Same crop in succession

Different crops simultaneously but with different maturation dates

- Make the most of one season with succession planting -

Succession planting is another great way to make the most out of a small urban garden. If you live in a warm to temperate climate, you've got a growing season long enough to accommodate two fast-maturing crops. You can start with peas or beans in the early spring, and then put cucumber plants in the same space as the pea crop ends and the warmer part of summer begins. You can also plant the same crop in waves over the first month or so of the gardening season to have a continuous harvest all summer long.

Chapter 5: Planning for Continued Success

One of the things that's most important for the ongoing success of an urban garden is keeping the soil healthy. If you only have a few small pots, replacing the potting soil each year will not be a major expense. But if you've got raised beds or an entire patio's worth of containers, you want to be able to save that expense and re-use your soil. This means adding organic material and putting your garden to bed for the off-season.

Your easiest route to do this is to get some bagged compost and a big roll of burlap. If you have perennial herbs planted in your raised beds, cut them down to about two inches about the soil line, and bring the cuttings inside to dry and crumble for your spice cabinet. You should cut any other perennial flowers down to the soil level as well. Remove any other spent plant material from the bed, and spread a layer of compost over the whole bed. Then roll out some burlap and stake it down, and you've tucked your bed in for the winter! When spring comes, turn the soil over to start the preparation process all over again. Remember, you can never have too much organic matter in your garden.

For your containers, you can do one of a couple of things. You can mix some compost into all of your pots and give them a little burlap lid for the winter- just cut a square, lay it over the pot, and tie a bit of twine around the outside to secure it-, or you can bag your soil in big contractor bags, layering in compost as you go. Then tie the bags up with as little air in them as possible, and set them someplace safe for the winter. Just pop the soil back in your pots and get back to business in the spring.

You can spend the off-season going through your journal notes and making plans for the following year. Check out seed and plant catalogs and decide if you want to try any different varieties or methods, and take a gander at gadgets made for small-scale gardening. You can jot down anything that piques your interest and snag deals on off-season sales and pre-sales. The research and the new toys are both parts of the fun of gardening, so enjoy!

Chapter 6: Community Gardening Considerations

One last note about urban gardening pertains to having a community garden. If you are interested in having a larger garden with your neighbors or building mates, you should check with the proper authorities to make sure you've got permission to use a shared space or community property. Community gardens are a great way to make sure that everyone has fresh food and beautiful flowers to brighten their lives all growing season.

Once you've made sure that it's okay to have a community garden, you can set some protocols in place to ensure that everyone knows the rules. Have a list of 'members' and divvy up time slots to make sure the work gets done. You can divide tasks up by days or weeks, and put a system in place so people get their fair share of the harvest. Have fun making friends and building a sense of community spirit! If you've got an overabundance, don't forget to share with a local food pantry or soup kitchen, to spread the love a little farther.

PART V

Chapter 1: What in The World Is A Thyroid

In this chapter, we will be exploring just what the thyroid is, and its functions in maintaining and keeping your body healthy.

For starters, the word thyroid has its origins dating as far back as the 1690s. Coming from the Greek, *thyreoiedes*, meaning "shield-shaped". Also referred to as, *khondros thyreoiedes,* "shield-shaped cartilage". Aptly named, as the thyroid gland is an organ which is often considered to resemble a butterfly, bow tie, or shield shape, at the base of the neck.

The thyroid gland plays an extremely vital role in the way your body uses energy, by releasing hormones that aid in controlling your body's metabolism. The hormones released by the thyroid gland assist in regulating important body functions, including but not limited to:

- Regulation of breathing
- Heart rate
- The central and peripheral nervous systems
- Body weight
- Cycles of menstruation
- The strength of muscles
- Levels of cholesterol
- The temperature of the body

Quite a lot for an organ coming in at only around 2 inches long. This tiny but important gland rests in the front portion of the throat, in front of the trachea, and just below the thyroid cartilage commonly referred to as the Adam's apple. The Adam's apple itself is the largest cartilage of the voice box or larynx.

Owing to the bow tie, butterfly, or shield shape, of the thyroid gland, is a middle connection of thin thyroid tissue, which is known as the isthmus, which is responsible for holding together two lobes on the right and left sides of it. It is not entirely uncommon, however, for someone to be missing the isthmus all-together and instead of having the two lobes of the thyroid gland operating separate from one another.

Now that you are aware of this and may feel a bit more familiar with your own thyroid gland, you may resist trying to see or feel around for it in your neck yourself. Unless the thyroid gland is otherwise afflicted and made to become enlarged, mostly known as a goiter, the thyroid will be unable to be seen, and only just barely able to be felt. It is only when a goiter occurs and the neck is swollen from an enlarged thyroid that it will be at all noticeable either to the eye or to the touch.

The thyroid gland is one of the major players in the endocrine system. The endocrine system includes glands that are responsible for the production and for the secretion of various hormones. The other organs which help make up the endocrine system are the hypothalamus, which is responsible for linking the body's nervous system to the endocrine system via the use of the pituitary gland. The pituitary gland which is responsible for the secretion of hormones, not the blood stream. The pineal gland, which produces the wake/sleep pattern hormone of melatonin. The adrenal glands, which are responsible for the production of a variety of hormones like the steroids cortisol and aldosterone as well as our body's adrenaline. The Pancreas, an organ located in the abdominal region of the body, the primary role of which is the converting of food into fuel for the body's cells. The ovaries and testicles, sex organs of the body. And the parathyroid glands.

Utilizing the iodine content from foods, the thyroid is able to produce and churn out the hormones T3, which stands for triiodothyronine, and the hormone T4, thyroxine.

T3, or triiodothyronine, is merely the active form of the companion hormone thyroxine, or T4. The thyroid gland alone is able to secrete around 20% of our body's T3 into the bloodstream on its own. With the other 80% coming from organs like the liver and the kidneys going through the process of converting thyroxine into its active counterpart.

It is absolutely possible for your body to have far too much of T3 though. When there is an over secretion of T3 into the blood stream, it is called thyrotoxicosis. This can be due to a number of conditions dealing with the thyroid gland such as overactivity in the thyroid gland, known as hyperthyroidism, caused by such conditions as a benign tumor, the thyroid gland becoming enflamed, or a condition known as graves' disease. The previously mentioned condition of a goiter, in which the neck begins to swell, might be a signal of thyrotoxicosis having occurred. Even more symptoms to have an eye out for in case of hyperthyroidism will be an increase in the appetite, increased regularity of bowel movements, an intolerance to heat, the loss of weight, the menstrual cycle becoming irregulated, a heartbeat becoming increasingly rapid or irregular in rhythm, the thinning or loss of hair, tremors, becoming irritable, overly tired, palpitations, and the eyelids retracting.

It is also possible for your body to be producing too little of the hormone T3. The thyroid gland producing too little of T3 is known commonly as hypothyroidism. It is common for autoimmune diseases to have a strong role in this occurring, an example of which would be the Hashimoto's disease, which causes the immune system to attack the thyroid gland. Certain medications or the intake of too little iodine can also cause hypothyroidism. This can be very serious, especially if a case of hypothyroidism goes unnoticed or untreated during early childhood, or even before birth. With the regulation of hormones being so important, primarily to physical and mental development, not treating hypothyroidism during these crucial times often result in reduced growth for the child, or becoming learning disabled.

The affliction of hypothyroidism is not foreign to adults though. When hypothyroidism occurs in adults they tend to have the functions of their bodies slowed down drastically. The effects of hypothyroidism in an adult have been known to include symptoms such as a growing intolerance to colder temperature, the heart rate of the adult will lower, gaining weight, a reduction in appetite, the ability of memory becomes poorer, fertility will reduce, muscles will become stiff, the adult may become depressed, and tired.

T4, or thyroxine, is the primary hormone that gets secreted from the thyroid gland and into the body's bloodstream. Unlike T3 which is active, thyroxine is in an inactive form and most of it will need to be converted to the active form, triiodothyronine, which is a process that takes place in organs like the kidneys and liver. Undergoing these processes is vital in making sure the body is able to regulate a healthy metabolic rate, control of the body's muscles, development of the brain, develop and maintain bones, and digestive and heart functionality.

As with T3, triiodothyronine, the production and secretion of too much will inevitably result in thyrotoxicosis, while the production and secretion of too little thyroxine, will result in hypothyroidism.

To combat this, the body and thyroid gland have a few tricks vital to the regulation of levels of these hormones in the cells. There is a controlled feedback loop system, involving the hypothalamus in the brain as well as in the thyroid gland and pituitary gland which is in control of the production of both of the hormones thyroxine and triiodothyronine. Thyrotropin-releasing hormones are secreted from the hypothalamus and, in turn, the pituitary gland becomes stimulated into producing thyroid stimulating hormone. A hormone which will stimulate thyroxine and triiodothyronine to be produced and secreted by the thyroid gland.

A feedback loop regulates this production system, to account for the levels of thyroxine and of triiodothyronine. If the levels of either of these thyroid gland hormones begin to increase, they will end up preventing the production and secretion of the thyrotropin-releasing hormone as well as the thyroid stimulating hormone, thus allowing the body to maintain, on it's own, a steady level of the thyroid hormones that it needs.

For all these reasons it is of vital importance that the levels of T3 and T4 being secreted thru-ought the body and its cells never get too high or too low. T3 and T4 are able to reach just about ever cell in the body by utilizing the bloodstream. The rate of work for the cells and metabolism to work is regulated by the hormones T3 and T4. To make sure that levels are never either too high or too low, this is why we have a thyroid gland.

The final hormone that the thyroid gland is responsible for the production of is the hormone calcitonin, CT, or thyrocalcitonin. Within the thyroid gland are what are known as C-cells, or parafollicular cells, which are in charge of the proliferation of this particular hormone. The primary role of calcitonin in the body is to help in the regulation of the levels of phosphate in the blood, and of calcium in the blood. Doing so is to be in opposition of the parathyroid hormone. In short, meaning that what it aims to do is reduce the amount of calcium in the blood stream. The reason for playing this role in the human anatomy game has been a bit of a mystery to science up to this point though, due to the observation of patients showing either very high or even very low levels of the hormone calcitonin, having no adverse effect on them.

The hormone calcitonin has two primary mechanisms by which to aid in the reduction of calcium levels within the human body. It can completely inhibit the activity of the cells in our body which are responsible for breaking down bones, known as osteoclasts. Osteoclasts do this because when bone is broken down, the calcium within the bone being broken down will be released into the body's bloodstream. So by inhibiting the osteoclasts from doing their respective jobs, calcitonin is directly involved in the reduction of the amount of calcium that is getting released into the body's bloodstream. Despite doing this though, the length of time that calcitonin can cause this inhibition has been shown to be quite short. Calcitonin can be an active player in the resorption of calcium into the kidneys, which it does by lower the levels of blood calcium in the body.

Calcitonin has been manufactured in the past and has then been given, in this form, to treat the disease of bone, Paget's disease. Also known as osteitis deformans, Paget's disease is rather common, and is a chronic bone disorder which can cause pain, fractures or deformities of a bone, or show absolutely no symptoms at all. It is however easily able to be controlled and treated with proper early enough diagnosis and treatment.

The manufactured hormone calcitonin has also been given to sufferers of general bone pain, and of hypercalcaemia, which is when the body has an abnormal level of calcium flowing in the bloodstream.

Though because of the introduction of bisphosphonates, which aid in the preventing of the breakdown of bone cells and are drugs also used to help treat osteoporosis, the use of manufactured calcitonin has decreased.

Chapter 2: Possible Thyroid Disorders

In the previous chapter, we began to cover what it is exactly that the thyroid gland gets done and even dabbled a bit into how it does it's job properly. During the last chapter, we mentioned a few of the various thigs which can afflict the thyroid gland, why this may occur in certain circumstances, and what the effects of these afflictions could be. Moving on into chapter two is where we will begin to take a closer look at everything that can go wrong with the thyroid gland. Not just the what, but the why as well. What causes these changes in our thyroid gland to occur, and what to expect to happen when they do occur. The importance of having this knowledge be a part of your thyroid gland arsenal cannot be at all overstated as there is a wide array of severity to both the symptoms and to the results of the ailments that can afflict the thyroid gland and consequently hinder our body's ability to maintain its health properly.

Just as well in this chapter, you can expect to be reading deeper into some of the ailments that may have already been brought up in the previous chapter, such as hyperthyroidism, hypothyroidism, graves disease, goiters, and Hashimoto's disease.

Hyperthyroidism

As briefly discussed in the last chapter, hyperthyroidism is a rather common condition in which there is overactivity in the thyroid gland and begins to produce far too much of the thyroid hormone which would usually be used to regulate the body's metabolic rate. This can be an overproduction of the hormones T3, which is triiodothyronine, T4, which is tetraiodothyronine, or even an overproduction of both of these hormones.

The causes of hyperthyroidism can vary greatly, with the most common reason for it being the aforementioned Grave's disease, which we will go much further into later in the chapter. The basics of Grave's disease are that it is an autoimmune disorder which causes antibodies in the body to stimulate the thyroid gland making it secrete to many of it's hormones. You should tell your regular doctor if any one in your family has ever had Grave's disease as it seems to have a genetic link, being passed down commonly from generation to the next generation. Grave's disease is also known to be more prevalent in women, affecting about 1 percent of the female population, than it is in men.

Another common reason for hyperthyroidism to occur is an excess level of iodine in the body, which is the main ingredient in hormones T3 and T4.

Less common, but still just as relevant to the conversation as causes for hyperthyroidism is thyroiditis which is the inflammation of thyroid gland, which in turn will cause the hormones T3 and T4 to start leaking out of the thyroid gland.

Tumors located on the ovaries or testes have been known links to hyperthyroidism. As well as even tumors, even when benign, located on the thyroid gland, or pituitary gland.

An easily preventable cause of hyperthyroidism which should not be overlooked is the intake of large amounts of T4, or tetraiodothyronine, via the ingestion of a dietary supplement or of a prescribed medication.

When it comes to the symptoms of hyperthyroidism, believe it or not, we had only scratched the surface in the previous chapter and will be going more in-depth here on what you can expect to look out for in order to self-diagnose an issue before going to seek out a professional opinion.

To begin with, in the case of Grave's disease, one of the symptoms can be a bulging of the eyes as if stuck in a stare. Other symptoms to watch out for would be an increase in the appetite, perhaps an increase in nervousness or a sense of restlessness. Muscular weakness, the inability to concentrate on simple tasks, irregularity in the heartbeat, loss of the ability to sleep soundly or for long periods of time, the loss of hair, or noticing that your hair has become thinner or more brittle, can be signs of hyperthyroidism. Thinness of the skin is also common, as well as becoming more irritable, sweating more, or becoming more anxious. In men specifically, the development of breasts can be a sign of hyperthyroidism. And in women, hyperthyroidism has been known to have adverse effects on the regularity of the menstrual cycle.

If you experience any of the prior symptoms, it is, of course, recommended to seek out professional help and diagnosis. However, it is highly recommended that you seek out professional help for the treatment of hyperthyroidism if you begin to experience a sensation of dizziness if you start to notice shortness in your breathing, which will likely come with the increase in heart rate, making it faster and irregular, and any loss of consciousness. Having hyperthyroidism has also been known to be the cause of atrial fibrillations, which are a dangerous arrythmia, commonly responsible for leading to having a stroke, or even to congestive heart failures.

In diagnosing a case of hyperthyroidism, a doctor will likely begin the process by conducting a full and complete medical history, as well as a physical exam. These are commonly conducted as they are helpful in revealing the common signs of loss of weight, how rapid your pulse is, an elevation in pressure of the blood, protrusion of the eyes, or the enlargement of the thyroid gland itself.

It is also reasonable to expect your doctor to conduct a cholesterol test which will be done to check on the levels of cholesterol in your system. This is done because cholesterol levels being low can be an indication that there is an elevation in your metabolic rate, which would mean that your body is burning through your cholesterol far too quickly.

Doctors are also able to conduct tests to measure the levels of T3 and T4 that are in your blood. Thyroid stimulating hormone tests can be done to check the levels of TSH, or thyroid stimulating hormone coursing within your body. TSH stimulates your thyroid gland to produce the hormones the body needs, and if your thyroid gland is producing levels of hormones at a normal rate, or even a rate that is too high, your TSH should come out lower. And a level of TSH that is abnormally low can be an important signifier that you may have hyperthyroidism.

A triglyceride test will be done, because similarly to having low amounts of cholesterol, a low level of triglycerides can be significant of an elevation in your metabolic rate. A thyroid scan or uptake will allow a doctor to see if your thyroid gland is being overactive. It will actually get even more particular, and let a doctor be able to see if it is the entire thyroid gland which is acting up or just a particular area of the thyroid gland.

Ultrasounds have been known to be utilized, as they will allow a doctor to observe entirely, the size of the thyroid gland, as well as any masses that may be within the thyroid gland. It is the use of the ultrasound which will also be able to let the doctor know if the mass inside the thyroid gland is cystic, or if it is solid. Just as well a CT, Computed Tomography, or MRI, Magnetic Resonance Imaging, scan can be performed to show if the condition is being caused by a tumor being present on the pituitary gland.

Treatment of hyperthyroidism also comes in varieties and may be dependent on the cause of the hyperthyroidism. Perhaps the most common treatment comes in the form of medication. Generally an antithyroid medication like methimazole, also known as Tapazole, which will cause the thyroid gland to halt the production and secretion of hormones altogether.

According to the American Thyroid Association, around 70 percent of U.S. adults who undergo treatment for hyperthyroidism will receive a form of treatment called radioactive iodine. Radioactive iodine is essentially able to completely and effectively destroy the cells that would otherwise be producing hormones. Radioactive iodine, or RAI, in the form of a liquid or a pill, will be ingested by way of the mouth, and is safe to use on an individual who has had any allergic reaction to an X-ray contrast agent or to seafood, because essentially the reaction comes from the compound which contains iodine, and not from the iodine itself. The iodine, in an iodide form, is actually split into two forms or radioactive iodine, known as I-123, which is harmless to thyroid gland cells, and I-131, which is responsible for the destruction of thyroid gland cells. The radiation which is emitted by both of these forms of the iodine are able to be detected from outside of the patient, which will help the doctor to gain any information needed the thyroid glands functionality, and take any pictures needed of the size thyroid glands tissues, as well as their location in the body. This treatment is not without its side effects though, which generally tend to come in the presence of dryness of the mouth, soreness of the eyes and in the throat, and has also been known to effect changes in taste. You may also be required, if undergoing this treatment, to take precautions for a short time which will prevent the spread of radiation to others.

Surgery is yet another common form of treatment for hyperthyroidism. In this case, it is entirely possible that a section of your thyroid gland will be removed, though entire thyroid glands have also been removed in this procedure. This is followed up with taking thyroid hormone supplements which will help in the prevention of hypothyroidism, which is what happens when there is the occurrence of underactivity in the thyroid gland, causing it to produce and secrete too little of the intended hormones. Beta-blockers may also be taken, such as something like propranolol to help control a rapid pulse, sweating, any anxiety that may crop up, and higher blood pressure. It is reported that most people respond very well to this form of treatment.

If you would like to improve any symptoms, or even take action to prevent symptoms from occurring, you are not left without options. You can work along with your doctor, or a dietician, to help create a healthy guideline for diet, exercise, and any nutritional supplementation. Proper diet intake, with a stronger focus on getting calcium and sodium, can be crucially important in the prevention of hyperthyroidism. Osteoporosis is a common result of hyperthyroidism as it can make your bones become thin, weak, and very brittle. To strengthen the bones after treatment for hyperthyroidism, it is recommended to take calcium supplements and vitamin D. To get an idea of how much vitamin D you should be taking post-surgery, you can talk to your doctor for a recommendation.

Moving on from treatment, it is not unusual for a doctor to recommend their patients to an endocrinologist, who will be more specialized in the treatment of systems dealing with bodily hormones. You'll want to avoid stress at this stage as it can cause thyroid storm, which happens when a large amount of thyroid hormone gets released, resulting in a horrible and sudden worsening of any prior symptoms. Proper treatment is both recommended and effective at the prevention of thyroid storm, as well as other complications such as thyrotoxicosis.

In the long-term, the outlook for something like hyperthyroidism is dependent heavily on what is causing it. Some of the causes of hyperthyroidism can go away without ever seeking treatment. Whereas a more serious cause like Graves' disease is not to be taken lightly, as it will get much worse if it goes without treatment, and the complications due to Graves' disease are often life-threatening and will have an affect on your quality of life long-term. These are easy enough to subdue with proper care and an early diagnosis and treatment.

Hypothyroidism

Though we went over a little about hypothyroidism in chapter 1, it is important to take a closer look at the disorder, to gain a better idea of its symptoms, and proper treatment and care for it.

When the body is not producing enough of the thyroid hormones that it needs, this is what is known as hypothyroidism having occurred. This will cause the general functions of your body to become slowed down, as the thyroid gland is responsible for producing and secreting hormones which will provide energy to nearly every other portion of your body. Though this affliction can come to task at any age, it is more common for an underactive thyroid gland to be noticed in adults over the age of 60, as well as being more prevalent in women. A diagnosis of hypothyroidism is nothing to get too worked up about, fortunately, as treatment of hypothyroidism has been known to be quite effective, as well as being very simple and very safe.

Though the symptoms of having an underactive thyroid gland can vary from person to person, there is enough overlap in the symptoms for us to help lay out what to look out for. It is important to note, however, that there can be difficulty in pin-pointing that a symptom is that of hypothyroidism and that the severity of the condition itself plays a large role in which signs or symptoms will appear, as well as when they may make an appearance.

It is not at all uncommon for most people to experience the symptoms of this condition arriving in a slow progression over many years. The thyroid gland will grow ever slower and slower, which will only then allow the symptoms to be better identifiable. The trouble can become that many of the symptoms come with general aging, so if you suspect there is more to the picture, and that hypothyroidism is at play, it is important to go see a doctor. An example of some early symptoms which also come naturally with age are the symptomatic fatigue and gaining of weight.

If hypothyroidism does occur, however, other symptoms to keep an eye out for will be an uptick in depression, constipation, or muscle weakness. It is also common to begin becoming more sensitive to the cold, for the skin to become dry, and a reduction in sweating. Your heart rate will generally become slower, blood cholesterol may elevate, and joints may become stiff or experience more pain. It is also possible for memory to start becoming impaired, hair may thin or become dry. Your voice may become hoarse, muscles will stiffen and experience soreness, your face will become puffy and sensitive. In women, hypothyroidism as been known to negatively affect menstrual changes and cause difficulty in fertility.

When it comes to the causes of hypothyroidism, an autoimmune disease is fairly common to be the culprit at work. The body is designed in such a way that your immune system generally will protect the body's cells against any invading bacteria and virus. Therefore, when an unknown virus or bacteria enters the body, it is the immune system which will respond by sending out what are known as fighter cells, to destroy the foreign invading virus or bacteria.

However, it is not impossible for your body to begin confusing what are the healthy and normal cells, with the invading cells. This is what is then called an autoimmune response to the cells. And if this autoimmune response does not get properly treated, or if it is not properly regulated, it is your own immune system which will start to attack your healthy body tissues. Medically, this has been known to cause quite serious issues, which include hypothyroidism.

Hashimoto's disease, which we have mentioned before, is one such autoimmune condition that can occur, and it is the most common among the causes of having an underactive thyroid gland. The disease literally will attack the thyroid gland which will cause chronic thyroid inflammation, which, in turn, will reduce the functionality of the thyroid gland. As with Graves' disorder having links between generations, it is not at all uncommon to find that multiple members of a family have this same condition as well.

Hypothyroidism can even become an occurrence as a result of treatment for hyperthyroidism, which has the aim of lowering your thyroid hormone. It is not uncommon for the treatment to result in keeping the thyroid hormone too low, which then becomes hypothyroidism, which has been a known result of the radioactive iodine treatment for hyperthyroidism.

The surgical removal of the thyroid gland is yet another known cause of the occurrence of hypothyroidism. The entirety of the thyroid gland will be removed in the case of thyroid problems cropping up, which will affect the body's ability to produce thyroid hormone, and cause hypothyroidism. In this instance, you will typically be recommended to take thyroid medication for the rest of your life. In the case that it is only a smaller portion of the thyroid gland which is removed, it is possible for the thyroid gland to still be able to produce and secrete a healthy amount of hormones. In which case it will take a test of the blood to determine how much medication you will need.

It is possible for radiation therapy to be the cause if you have come down with hypothyroidism. A diagnosis of leukemia, neck cancer, or lymphoma will likely mean you have had to undergo a form of radiation therapy, which very nearly almost leads to the occurrence of hypothyroidism.

Just as possible, is a medication you may be taking to lower thyroid gland hormone production, to be the cause of hypothyroidism. Medications such as these are commonly used in the treatment of certain psychological diseases, and even have been known to be used in treating heart disease and cancer.

When it comes down the diagnosing of hypothyroidism, there are two primary methods which have been favored and work to best identify when it has occurred. The first being a strict medical evaluation, much like in the case of checking for hyperthyroidism. The doctor will give you a very thorough exam physically, as well as making sure to go over your medical history. Hypothyroidism has a couple physical signs which the doctor will be checking for primarily such as the dryness of the skin, how slow or quick your reflexes are, any swelling of the neck, and the rate of your heart beat. It is at this time that a doctor will also likely ask you to report any of the other symptoms listed earlier that you may have experienced, such as the depression, any fatigue, if you have been constipated, and a sensation of being more sensitive to the cold. It is also at this point it will be most helpful for you to let the doctor know of any thyroid conditions which have existed in your family.

To reliably get an idea of the existence of hypothyroidism in the body, it is required to conduct blood tests. It is only by this method that anyone will be able to tell and get a look at a measure of your body's thyroid-stimulating hormone levels, done by utilizing a thyroid-stimulating hormone test to see how much of the thyroid-stimulating hormone your pituitary gland is or is not creating. In the case that your thyroid gland is not producing enough of the hormone, the pituitary gland will respond to this by boosting the thyroid-stimulating hormone it produces in order to increase thyroid hormone production. If it turns out you have hypothyroidism, the levels of thyroid-stimulating hormone in your body will be increased, because your body is responding by making an attempt at stimulating more thyroid gland hormone activity. If hyperthyroidism is what ails you, the levels of the thyroid-stimulating hormone in your body will as having decreased, because in this case, your body has begun the process of attempting to halt the function of excessive production of the thyroid glands hormones.

Another useful method in the detection and diagnosis of hypothyroidism is to test the levels of T4 in the body, being produced by your thyroid gland, as T4 is produced directly by the thyroid gland. When they are used in conjunction with one another, a test of T4 levels and the thyroid-stimulating hormone test are very helpful in coming up with an evaluation of thyroid gland functionality. In general, if you the levels of thyroid-stimulating hormone in your body has increased, while the level of the hormone T4 has decreased, you much more than likely have hypothyroidism. Though, due to the sheer amount of conditions that can have such a negative impact on the thyroid gland, it could very well end up being necessary to conduct even more tests of the thyroid glands function I order to properly diagnose the issue.

Though it is true that for many people who have thyroid conditions, that the right amount of the proper medication will assist in the alleviation of their symptoms, you will have hypothyroidism for the rest of your life if you get it.

To get the best of hypothyroidism it is most commonly treated the best with the use of levothyroxine, also known as Levothroid or Levoxyl, which is T4 put into a synthetic form that is responsible for copying the action the thyroid hormone would regularly take if it were being produced as normal by your body. The idea behind doing this is that the medication will cause a return to the proper levels of the thyroid hormone in your blood. Once a restoration of the thyroid hormone level has occurred, many of the symptoms that come along with having hypothyroidism, will at the very least become much easier to manage, and at best the hypothyroidism symptoms will disappear altogether. It is important to expect it to take several weeks, following treatment, before relief sets in, and you start to feel a return to normalcy. There will also very likely be follow up appointments for testing your blood, which the doctor will recommend in order to keep a solid eye on your progress into recovery. Chances are that you will also receive some medication or other recommended methods to aid you in your recovery, be sure to speak with your doctor about the dosage you should be taking and to come up with a solid plan, that will most benefit you, for recovering in a timely fashion.

It is the case that many people who end up with hypothyroidism medicate for it, for the rest of their lives. Despite this, the dosage you will be taking thru ought that time is likely to go through changes. To better get an idea of how these dosages should be changing over time, it is best to get a check up on your thyroid-stimulation hormone levels every year. In this way, your doctor will be able to more properly adjust the amount you should be taking, or not taking, based on the blood levels indicated by the thyroid-stimulating hormone tests. Only by doing this regularly, will you and your doctor be able to achieve the recovery program that works best for you.

Plans and programs for this achievement may include medications and other hormone supplementation. Once again, synthetic versions of the hormone you need may be used, as they are a widely used and viable practice to aid in the recovery of hypothyroidism. The synthetic version of the hormone T3 is liothyronine, and T4 in its synthetic medication form is called levothyroxine, both of which act as suitable substitutes for their corresponding hormone.

If it was a deficiency in your iodine intake which caused your specific occurrence of hypothyroidism, it is likely that your doctor will recommend a supplementary form of iodine. Keep in mind to ask your doctor, and get the proper testing before taking anything, but selenium and magnesium supplements have been known to aid heavily in the treatment of hypothyroidism.

The golden ticket to any recovery or treatment is usually diet, and in the case of hypothyroidism, there is no exception. Though this is the case, and diet can be incredibly beneficial in your recovery and treatment, do not expect a change in your diet, doctor recommended or otherwise, to replace the need for a prescribed medication. Foods that are rich in selenium or magnesium such as nuts and seeds like the Brazil nut and sunflower seeds have been shown to be very beneficial additions to any diet to aid in the treatment of hypothyroidism.

Balance in your diet will play an especially important role, as the thyroid gland requires particular amounts of iodine in order to properly reach full functionality. There are foods such as whole grains, vegetables, fruits, and lean meats which can handily accomplish this without the need for iodine supplementation.

And of course, diet is only the beginning, exercise as well comes in as an important slice of the treatment and recovery pie. The muscle and joint pain that coincides with hypothyroidism will more often than not leave one to feel extreme fatigue and depression, both of which can be helped by creating and sticking to a regular work out regime. Though no exercise should be discounted, unless specifically told to avoid certain activities by your doctor, there are certain ones which will prove more beneficial than others for treating the symptoms of hypothyroidism. Low impact workouts such as swimming, riding a bike, doing Pilates or yoga, or even a good brisk walk, have been known to be very helpful low impact work outs that are helpful and easy to work in to a daily routine.

The building up of muscle mass by strength training, lifting weights, sit ups, pushups, and pull-ups, help reduce the lethargic feeling of sluggishness that comes along with hypothyroidism. The increase in muscle mass will result in an increase in the rate of your metabolism, which will simultaneously assist in decreasing any weight gain that the hypothyroidism may have caused.

And finally doing training that is primarily cardiovascular. As stated earlier, hypothyroidism is one of the ailments that can correlate with a heightened risk of having a cardiac arrythmia, or irregularity of the heartbeat. By taking steps to be more mindful of your cardiovascular health, exercising on a regular basis or schedule, will help in protecting your heart.

There are also alternative treatments which exist to help in taking care of hypothyroidism, such as animal extracts that contain the thyroid hormone. These extracts are made available from pigs because they contain both the thyroid hormone T4 and thyroid hormone T3. It is uncommon for these to be recommended, however, as they have not shown to be reliable in how to dose, as well as not being more effective than the typically recommended medications. It is also popular to find some glandular extracts in stores that are health food based. The risk that comes along with them is that the U.S. Food and Drug Administration plays no role in the monitoring or the regulation of these extracts. This has historically brought the guarantee of their pureness, legitimacy, and even their potency into question. If you decide to use these products, you do so at your own risk, but still be sure to inform your doctor so that they can adjust accordingly to your treatment.

You can go above and beyond in regards to hypothyroidism treatment, yet still deal with issues or complications that are longer lasting because of this harsh fluctuation to your body. Luckily there have been methods developed and used which will help to lessen the burden of hypothyroidisms effects on your life moving forward.

In the beginning, fatigue can feel like a lot to deal with, especially when associated with depression. These feelings can creep through even if you are taking proper dosages of your medication. It is of utmost importance that you get a good quality sleep every night to ease your treatment and recovery. A good, healthy diet, as well as the relief of stress through activities such as meditation, Pilates, and yoga, are effective strategies when it comes to combating lower energy levels.

It is also vitally important to recognize the difficulty of having a medical condition that is chronic, especially in the case of something like hypothyroidism, which comes along with its own mixed bag of other concerns to your overall health. Being able to talk about, or express, the experience of going through this will help. There are resources out there for support groups of other people who live with the effects of hypothyroidism, you can find a therapist to talk to, perhaps a close friend or loved one. Anyone who will be able to enable you to discuss your experience with openness and with honesty. You may even be able to receive a recommendation for meetings of people with hypothyroidism, from an education office at your local hospital. Connecting and communicating with others who can empathize with what you are going through could end up being an enormous aid in your recovery and life with hypothyroidism.

Important as well is making sure you monitor yourself for other health conditions that could arise. As we went over earlier, the main cause for hypothyroidism is an autoimmune disease. Just as well, links with hypothyroidism have also been found in conditions such as diabetes, having pituitary issues, having your sleep obstructed by sleep apnea, and lupus.

Just as with fatigue, depression is a common symptom and side effect of going through and living with hypothyroidism and should be watched closely. The thyroid glands hormone levels lower, the function of your body begins to slow down, and before you may realize it you are living with a depression that was not there before. It is vital to know what to look out for, and not just what, but also how to look after yourself while dealing with this.

Depression as a symptom can make hypothyroidism difficult to diagnose as there are many who may only experience difficulties or changes in mood as a symptom. It is for this reason, that instead of having a doctor check only your brain when checking for depression, it can also be important to ask them to check for signs of you having an underactive thyroid. Aside from the changes in mood, there are a few other similarities that exist in both having depression as well as hypothyroidism such as, gaining weight, finding it difficult to maintain concentration, feelings of daily fatigue, which coincide with a reduced desire and satisfaction with daily life, and hypothyroidism or depression could both effect your ability to sleep well.

Not all of their symptoms overlap so nicely though, both have their conditions which differentiate one from the other. In the case of hypothyroidism there are, of course, some physical signs such as the dryness of the skin, or the thinning and loss of hair. There is also the tendency to become constipated and the increase in levels of cholesterol.

These symptoms would be atypical if depression alone was the issue.

If you have hypothyroidism and it is the cause of your depression, then the correct treatment and care of the hypothyroidism should be just the remedy needed in order to treat your depression as well. If the hypothyroidism passes and depression remains, it may be important to talk to your doctor about receiving further help and a change in medication.

Along with depression being a symptom of hypothyroidism, it has recently been found, through studies, that around 60 percent or so of people who get hypothyroidism tend to also exhibit having anxiety as well. Studies are ongoing and are still growing in scope and size, though it would still be in your best interest to discuss all possibilities and symptoms with your doctor in order to more thoroughly and best tackle the treatment of hypothyroidism.

It cannot be stressed enough, how much of your body is under the affects and influence of your thyroid gland working properly to produce and secrete the correct levels of hormones. For this reason, when a woman gets hypothyroidism and simultaneously desires to get pregnant, she will be faced with her own subset of challenges to come. Have a low thyroid gland function during a pregnancy can cause a number of conflicts including various birth defects, have a still-birth or miscarriage, as well as anemia or a low birth weight. It is not uncommon for a woman with thyroid problems to have a perfectly healthy pregnancy, but to make sure that you reach this outcome it is important to do things such as eating well, keeping yourself informed about current and effective medicines, as well as talking to your doctor about testing.

Though testing may result in changes to your dosage or medication, it is also for this reason that it is important to make sure you are not deviating from the medications provided and the dosage your doctor has recommended.

Considering the thyroid issues adds on even more importance to the need for eating healthy while pregnant. Make sure that you are getting the proper amount of vitamins, minerals, and nutrients and consider taking multivitamins as well to supplement this.

It is not impossible to develop a thyroid issue such as hypothyroidism while pregnant. In fact, for every 1,000 pregnancies, this tends to occur in every 3 out of 5 women. It is important for doctors to routinely check thyroid levels during your pregnancy, as some will do, to make sure your thyroid levels aren't becoming to high or low. If they end up being higher or lower than they ought to be, it is likely that your doctor will recommend you starting treatment. Even some women who have never before had any thyroid issues may develop them once the baby is born, which is known as postpartum thyroiditis, and also tends to resolve itself after a year in around 80 percent of the women it shows up in. It is only the other 20 percent of women who will have this happen and then go on to require the long term treatment.

When hypothyroidism takes place, and the functions of the body slow down, it is quite typical for people to become prone to gaining weight, which is very likely due to what happens to the bodies ability to burn energy, which is that the efficiency to do so slows down as well. This change in the body will typically cause someone who has hypothyroidism to gain anywhere from 5 to 10 pounds in general, making the weight that is gained not entirely drastic, but someone could still find it quite alarming. It is very possible then, that once the hypothyroidism has been treated, that any weight gained will then be easily lost. If this does not occur, a simple change in diet, and adding regular exercise to your routine should aid in handily losing the weight, as your ability to manage weight will go back to normalcy, with the return to proper levels in your thyroid hormones.

Hypothyroidism is a common occurrence; therefore it is also commonly treated without issue. Hypothyroidism has been found to occur in around 4.6 percent of the American population that are 12 and older. Which comes out to about 10 million or so people who go on to live long healthy lives with the condition, and you may never even realize it. It is far more prevalent in people who are over the age of 60, and in women about 1 in 5 of them are likely to experience hypothyroidism by the time they have reached 60 years of age. One of the causes is Hashimoto's disease which happens to appear more in women who have reached middle-age, though it can absolutely show up in children and men. As Hashimoto's disease is hereditary, it is likely that if you get it, you did so from a relative, and have an increased chance then of passing it on down to your children.

It is important to keep an eye on your body, your health, and your thyroid gland as you get older. If, as the years go by, you begin to notice any of the changes gone over in this chapter so far, it is vital that you see a doctor in an attempt to get a proper diagnosis and seek treatment as soon as possible.

Hashimoto's Disease

Hashimoto's disease is an autoimmune disease which can be very destructive to your thyroid gland, and thereby your thyroid glands ability to function properly. Hashimoto's disease is also known as chronic autoimmune lymphocytic thyroiditis and is the most common cause of having an underactive thyroid gland, hypothyroidism, in the United States.

As an autoimmune disorder, Hashimoto's disease is one of many conditions that will be the cause of your body's white blood cells and your body's antibodies becoming confused and starting to attack the cells that make up the thyroid gland. What makes this happen precisely is still somewhat of a mystery to doctors, even still it is believed by some that factors of genetics may be involved.

With the cause of Hashimoto's disease being unknown, it is difficult to precisely put a finger on what puts a person at risk for having or contracting the disease. There are still, however, just a few factors that doctors are aware of which could signify being at risk for the disease. In the case of Hashimoto's disease, in particular, women happen to be seven times more likely to contract than men, and especially for women who have been pregnant before. Having a history of autoimmune diseases in the family is another factor that could mean you are at higher risk of having Hashimoto's at some point in your life, especially if the autoimmune diseases include Graves' disease, lupus, rheumatoid arthritis, if there is a history of Sjogren's syndrome in your family, or a history of type 1 diabetes, Addison's disease, and vitiligo. If it is the case that these autoimmune diseases are present in your family line or may have been based on symptoms of Hashimoto's disease, get together and discuss the possibility with your doctor, then make sure to get tested for the disease.

Hashimoto's disease is interesting in that the symptoms of it, are not symptomatic of Hashimoto's disease alone, in fact, they are similar to having the symptoms of an underactive thyroid gland, or hypothyroidism. Some signs to watch out for that your thyroid gland is not working properly to produce proper thyroid hormones, and that you may have Hashimoto's disease are your skin becoming dry and pale, constipation, if your voice becomes hoarse, you become depressed and start to feel sluggish or fatigued. High levels of cholesterol, a thinning of the hair, muscle weakness in the lower body, and intolerance to the cold may also be signs of hypothyroidism as a result of Hashimoto's disease. In women, it can also cause issues with fertility. Hashimoto's can exist inside of your body for many years before you begin to show any signs or symptoms, and during that time, it may progress while showing no signs of damage to the thyroid gland. Some with Hashimoto's disease end up with a goiter, an enlarging of the thyroid gland which causes the front of the neck to swell. Though generally painless, it is common for a goiter to make the act of swallowing difficult and for it to simulate a feeling of fullness in the throat.

Owing to it's difficulty to diagnose, your doctor may not suspect Hashimoto's of being prevalent until observing symptoms having hypothyroidism. In which case they will need to conduct a blood test designed to check the thyroid-stimulating hormone, or TSH, levels in your body. It is a relatively common and safe test, which is also an accurate way to check to see if you have Hashimoto's disease. Levels of thyroid-stimulating hormone are higher when the activity of the thyroid glad is lower because your body starts working harder to stimulate the production of more thyroid hormones to secrete from the thyroid gland. There are also blood tests that your doctor may conduct if they feel the need to check further for the levels of antibodies, cholesterol, and other thyroid hormones, T3 and T4, in your blood. Testing for all of these can help immensely in pinning down a diagnosis of Hashimoto's disease.

Unless your thyroid gland is functioning normally, in which case your doctor may still recommend regular checkups to monitor you for any changes, it is very likely that the need for treatment of Hashimoto's disease will be required.

The improper production of enough hormones in your body by your thyroid gland will likely result in the need to take medication. In the case of having to take this medication, it is also likely that you will be prescribed on it, though dose will vary, for the rest of your life. The effective drug most commonly prescribed is levothyroxine which is the hormone thyroxine, or T4, made synthetically, and which will successfully replace the missing hormone in your blood. The synthetic hormone drug levothyroxine tends not to have any noticeable side effects, and regular use has been known to frequently return the hormone levels of the body back to normal, restoring proper function of the thyroid gland. When this happens, all other symptoms of Hashimoto's disease and hypothyroidism generally tend to disappear altogether, though it is likely that your doctor will still recommend that you still get regular testing done so that your hormone levels can be consistently monitored to prevent something like hypothyroidism from becoming a problem again moving forward. Getting the regular testing allows the doctor to adjust the dosage of your medication as necessary if at all necessary.

It is important to consider before going on levothyroxine, that there are supplements and medications which will have an effect on your body's ability to absorb the drug. As such, make sure you have a discussion about this with your doctor if you are taking any other medications, especially if they include iron or calcium supplements, or estrogen. Some medications for cholesterol have been known to cause an issue, as well as proton pump inhibitors which are used as a treatment for acid reflux.

Though these have been known to cause an issue, there is what could be an easy work around of simply changing what time of the day you take your other medicines in conjunction with the doctor recommended thyroid medicine. It is also possible that certain foods could end up being involved in the efficacy of your thyroid medicine. It is best to discuss all of this with your doctor to come up with an efficient way for you to take your thyroid medicine, based on your dietary needs.

The severity of complications due to leaving Hashimoto's untreated varies and are not worth the risk if you ever contract, or if you have it. They go far beyond just hypothyroidism and include heart problems that an include total failure of the heart. It is not unusual for anemia to be a result of leaving Hashimoto's disease unattended. Depression and a decrease in libido are common, as well as higher levels of cholesterol in the blood and experiencing a sense of confusion or loss of consciousness.

Hashimoto's disease has also been the culprit responsible for complications during a woman's pregnancy cycle. It is far more likely, that if you carry out a pregnancy while having untreated Hashimoto's disease, that you may be putting your child at higher risk of being born with defects of their kidneys, their heart, and even their brains.

These complications can be limited by talking to your doctor during the pregnancy and keeping on top of monitoring your thyroid glands hormone levels with the proper blood testing. If you are a pregnant woman and have thyroid issues, such preventative measures could mean a severe change in the life and health of your child. However, if you have not had any known disorders with your thyroid or hormone levels, it is not recommended that you get regular or constant screening done during the pregnancy.

Graves' Disease

Another autoimmune disorder, Graves' disease is responsible for causing your thyroid gland too create too much of the thyroid hormones in your body. When this happens it is a condition referred to commonly as hyperthyroidism. Graves' disease, is named such for the man who discovered it, an Irish physician named Robert J. Graves, and is regarded as one of the most common forms hyperthyroidism takes, having an effect on around 1 out of every 200 people.

When Graves' disease occurs in the body, it will cause your immune system to begin creating antibodies that are known as thyroid-stimulating immunoglobulins, that attach themselves to the body's usually healthy cells of the thyroid gland. By doing this they end up causing the thyroid gland to produce and secrete more of the thyroid hormones than it is meant to for your body. The hormones that are produced by the thyroid gland go on to affect a great number of your body's functions including its temperature, the function of the nervous system, the development of the brain, and the list goes on. For this reason, hyperthyroidism can end up having a negatively driven affect on not just all of those functions, but when left untreated can also cause the loss of weight and mental and physical fatigue. Hyperthyroidism has also been found to be responsible for such things as depression and emotional liability where the individual will uncontrollably cry or laugh or put on other manic emotional displays.

Due to the role that Graves' disease can invariably play on the appearance of hyperthyroidism in the body, it is no surprise that the two would contain a sharing of many of the same symptoms. These symptoms include tremors especially of the hands, a loss of weight, tachycardia, which is the rapidity in the rate of the heart, becoming intolerant to heat or warmth, fatigue, nervousness and irritability, the swelling of the front of the neck, due to the enlargement of the thyroid gland, known as a goiter, an increase in the frequency of having bowel movements, as well as diarrhea, weakness of the muscles, and having it become difficult to get a good full night's worth of sleep. Among the people who experience having Graves' disease, it is only a small percentage who will experience the skin thickening around their shin area and become reddened, an affliction which is known as Graves' dermopathy.

Another common symptom of Graves' disease which one may go through while experiencing the condition, is what is called Graves' ophthalmopathy. Graves' ophthalmopathy is what occurs when the eyes of the afflicted individual appear to be enlarged, which is a result of the eyelids retracting. When Graves' ophthalmopathy happens, it is entirely possible that your eyes may begin to bulge outwards from your eye sockets.

Estimates say that as much as 30 percent of the people who end up developing Graves' disease will observe at least a mild case of what is known as Graves' ophthalmopathy and that for up to 5 percent of the people will instead experience an extreme case of the eye bulging.

Because of autoimmune diseases such as Graves' disease, the immune system will begin to fight against what are the healthy cells and healthy tissues of the body. Normally, your immune system is producing proteins which are known as antibodies, which are responsible for fighting against foreign invaders to your body, the likes of harmful viruses and bacteria. The antibodies produced this way are formed especially with the duty of targeting a specific invader to the host. When it comes to the effect of Graves' disease on the body, your immune system begins to mistake healthy thyroid cells as these foreign harmful cells and produces the thyroid-stimulating immunoglobulins which then mistakenly go off to attack what are your healthy thyroid cells.

Scientists and doctors alike, are aware that it is indeed possible for a person to have inherited the ability for their body to make antibodies which then go against their own healthy cells, yet they have made no determination that such an occurrence is what is the cause for Graves' disease, or who will end up developing Graves' disease.

Despite that though, there are experts who believe that they have been able to button down on some factors which may increase ones risk for the development of graves disease which includes its tendency to be hereditary. So be sure to discuss family medical history with your doctor and talk about whether or not there are family members who have, or who ay have had Graves' disease. It is also believed by these experts that stress, gender, and someone's age may be some of the facets that end up putting someone at higher risk of getting Graves' disease. It is typical for the disease to be found in people who are younger than the age of 40, and it has been more prevalent, about seven to eight times so, in women rather than men.

Having had, or having still, another autoimmune disease is yet another factor that will increase your risk of ever getting Graves' disease. Examples of such autoimmune diseases are having Crohn's disease, rheumatoid arthritis, and diabetes mellitus, among others.

For the diagnosing of Graves' disease, when it is suspected, it is not unheard of for your doctor to request lab tests. The use of your families medical history as well, especially if there is a case of someone in your family having had Graves' disease, will be able to help act as a basis for your doctor to zero in on diagnosing whether you have Graves' disease as well or not. This is something that thyroid gland blood tests will be needed for in order to confirm. Your doctor may request that these tests and others may be handled by a specialist expert in diseases which are related to the body's hormones, known as an endocrinologist, in order to help get the diagnosis of Graves' disease. Other tests which your doctor may request are full bloodwork tests, a thyroid gland scan, an uptake test utilizing radioactive iodine, a test for levels of TSH, or thyroid stimulating hormone, and a TSI test, which is the thyroid-stimulating immunoglobulins.

By combining the efforts of the endocrinologist, as well as the array of tests, it is more possible for your doctor to determine if you do indeed have and need treatment for Graves' disease specifically, or if another thyroid disorder is what is at work, and thus requires its own specific form of treatment.

There are a number of options available for treatment when someone is diagnosed as having Graves' disease. These are generally the taking of anti-thyroid drugs, therapy in the form of RAI, or radioactive iodine, and getting thyroid gland surgery. It is not abnormal for a doctor, in the case of Graves' disease, to recommend, all, two, or just one of the treatments for the afflicted.

When it comes to treatment via anti-thyroid drugs, you will typically be taking medications such as methimazole, which is taken orally as a tablet and works by putting a stop to the thyroid gland producing and secreting too much thyroid hormone, and propylthiouracil, which is also taken orally and generally used as a back up if a drug like methimazole did not end up working well enough. The use of beta-blockers is also recommended on occasion as they are used in assistance of reducing the effects of symptoms until another treatment method can start working.

It is radioactive iodine treatment, or RAI, which is among the most common treatments suggested to those suffering of Graves' disease. It is required, during this treatment, that the individual seeking treatment take specified doses of radioactive iodine-131, the purpose of which is to destroy thyroid cells. The radioactive iodine-131 will be ingested orally, in small amounts, via pill. Be sure to discuss with your doctor and risks or precautions that come with this treatment.

The less frequent option for treatment is the thyroid surgery. This treatment will tend to be a last resort if the other options have not worked to full capacity, if there is a reason to be suspect of thyroid cancer being present, or if the patient is a pregnant woman who is unable to take any of the regularly prescribed anti-thyroid drugs.

In the case of surgery being necessary, it is not uncommon the doctor to issue the removal of your thyroid gland completely, in the interest of preventing the return of the hyperthyroidism. In which case, thyroid hormone replacement surgery will be necessary on a regular basis. Talk to your doctor about the possible side effects of choosing to go through with surgery, as well as generally what to expect moving forward.

Goiter

A goiter, goitre, thyroid cyst, or Plummer's disease, is a general term used for when there is an observable enlargement of the thyroid gland, usually resulting in a noticeable swelling of the front of the neck. Treatment for a goiter can be handled in a variety of ways, and the treatment method is dependent on the goiters location, the length of its presence, and how exactly it is affecting the thyroid glands performance.

Though usually unable to be seen or even felt, the thyroid gland generally tends to become detectable by touch and even perceptible to the eye when there is a goiter present. An expanse of the thyroid gland, or goiter, can be the cause of the whole thyroid gland expanding, which is known as a "smooth goiter", or just a part of the thyroid gland expanding, which is also called a "cystic" or "nodular" goiter. A goiter is not a sure symptom of having an active thyroid, known as hyperthyroidism, or underactive thyroid, known as hypothyroidism, and, in fact, the majority of people who have a goiter, retain a perfectly normal use of their thyroid gland.

A number of reasons exist for the existence of a goiter. Among these are included a deficiency in your levels of iodine. Iodine may be a trace element, but it is far from trivial. It assists in helping the thyroid gland in maintaining proper functionality and making the thyroid glands hormones. There are two primary hormones which are produced and secreted by the thyroid gland, these are T4 or thyroxine, and T3, also known as triiodothyronine. The approximate number of people who have iodine deficiency comes out to about 2.2 billion and it is estimated that around 29 percent of the worlds total population live in an area that is considered to be deficient in iodine. It is reported that people in the U.K. have proper levels of iodine as a part of their regular diet. If you are keeping your eye out for food sources that are a good source of iodine, there are salts that have iodine supplements, as well, non-organic milk is plentiful with iodine.

Thyroiditis is anther well known cause of goiter. Thyroiditis is more commonly referred to as when the thyroid gland has become inflamed. Around the world, the most common reason for thyroiditis occurring is Hashimoto's disease, or Hashimoto's thyroiditis, which is an autoimmune disease that causes the bodies antibodies to start to become confused and begin attacking healthy cells of the thyroid gland. Hashimoto's disease is not the only cause of the thyroiditis condition though, it could also stem from viral infection, and has been known to occur just after or during pregnancy.

A goiter has also been known to occur due to Graves' disease, another autoimmune disease, this one causing the immune systems antibodies attacks on thyroid cells to make the thyroid gland overactive, resulting in hyperthyroidism. It is this hyperthyroidism, or over activity of the thyroid glands capacity for producing and secreting hormones, which is the cause of the swelling of the thyroid gland.

If there are benign growths on the thyroid gland, they have been known to cause a goiter, most commonly known for doing this is a follicular adenoma, which can be a firm or rubbery tumor surrounded by a fibrous capsule.

External factors that may be the cause of goiter are known as goitrogens. Included among what would be considered a goitrogen are medicines such as the mental health drug lithium, and cabbage type vegetables. Ingestion in the excess of these vegetables, which include cassava or kelp, will likely result in the growth formation of a goiter.

There are physiological demands put on the body during pregnancy and during puberty which have been known to be at the root of a goiter. And as with other causes like Graves' disease and Hashimoto's disease, there is a strong likelihood of inherited genetic reasons that one may at some point experience goiter.

Due to the varying reasons for the existence of goiter, there are also a multiplicity of types of goiter. The first of these types is known as colloid goiter, or endemic goiter, which is a development due directly to a lack of sufficient iodine levels. As a result, the people who tend to end up with a colloid goiter are those we mentioned, who live somewhere with a less dense supply of iodine.

The next type of goiter is the nontoxic goiter, or sporadic goiter, as it is also well known. Though the definite cause of a goiter of this type is regarded as generally unknown, it is surmised that a sporadic goiter is a result of taking medications, such as lithium, for example, or so it is believed. Among the may uses for lithium, it is perhaps most commonly recognized as the drug used for aiding in the treatment of mood based disorders, the likes of bipolar or depression. The nontoxic name is apt in regard to this form of goiter, as they are benign, and have no discernable effect on the production or secretion function of the thyroid gland, leaving the thyroid to function at a healthy and normal capacity.

The final type of commonly recognized goiter is known as the toxic nodular or multinodular goiter. Generally originating and taking form from as merely an extension from what was a simple goiter prior, the toxic nodular goiter will take the form of at least one, but often more, small nodules on the expanding thyroid gland. This toxic nodular goiter, having taken a sort of root on the thyroid gland, then begins to produce its own thyroid hormone, which plays a big part in the causation of hyperthyroidism.

As mentioned above, it can be difficult to detect goiter before it has really taken effect to the thyroid gland, but after it has begun doing it's work it is most common for it to cause a swelling of the front of the neck, making it clearly visible as well as felt. Before the expanding has commenced, it is common to have had nodules existing in your thyroid gland, these small nodules cannot be felt, and may have even been only a chance occurrence due to examinations, and of scans, that were applied for other reasons. Cases such as these are rather common, and when they occur, there has been a tendency to notice no sign of a goiter up to that point. As nodules appear on the thyroid, ranging from smaller nodules to much larger nodules, it is the presence of these nodules which is what is the cause of noticeable swelling of the neck.

This swelling and the nodules which are collecting on the thyroid gland cause other symptoms to occur, like having a difficult time of swallowing or of trying to breathe, it is not uncommon for coughing to be a symptom, your voice may start to become hoarse, and there may be a dizzy sensation that is noticeable when you raise an arm above your head.

Goiter is a rather common occurrence. It is calculated by the World Health Organization, that around the world, goiter affects nearly 12 percent of the global population. However, it has also been recorded that in Europe, the rate of goiter is lower by a slight amount. Goiter being considered endemic, or noticeably affecting a certain area is a common occurrence wherever iodine is scarce, and the endemic definitions are only applied when goiter is recognized on 1 out of 10 people within a certain population.

It is usual for goiter to be the diagnosis when there is noticeable swelling on the neck that can be seen without the need of a scan, also making it detectable with the touch of the hand, due to the enlarged thyroid gland in your neck, something a doctor will be quick to check for, likely before anything else.

There are also a number tests a general practitioner may order in order to examine the levels in your blood of thyroid hormones coming from the thyroid gland, as well as wanting to make sure of the levels of antibodies that are prevalent in the bloodstream. This examination will often take the form of blood tests, that are used to detect the changes in levels of the hormones as well as whether or not the level of production of the antibodies has increased, which tends to happen in response to the body experiencing an injury or infection in the blood.

A thyroid scan, or thyroid uptake scan, will show the size of the goiter itself, as well as what condition the goiter is in. It will also aid in identifying any differences in activity, in various places on the thyroid gland.

A biopsy may be recommended, the procedure of which involves removing samples of your thyroid gland, and then sending the samples of your thyroid gland's tissue to an outside laboratory or endocrinologist for examination.

It is also possible that an ultrasound scan may be used which will help for a doctor to see images of the inside of your neck, getting a much closer look at the size of the invasive goiter, allowing for the observation of nodules. As more ultrasounds are done, it is even then possible to track the changes in size or shape of the nodules, and the size of the goiter.

You may, at some point, be referred to an endocrinologist in order to get an outpatient assessment, giving you and the doctor more information from the examination by an expert. During their examination you may have to undergo a test known as a fine needle aspiration, which is done on the thyroid gland. For the procedure to take place, the endocrinologist will make use of a fine needle which, utilizing the guiding sight of an ultrasound, will be used to remove tissue from your thyroid gland, only a small amount will be needed. The tissue removed from your thyroid gland is then studied under the lens of a microscope, which will assist the endocrinologist in assessing exactly the types of cells which are currently present in your thyroid gland. It is not at all uncommon for a procedure like this to need to be repeated one or more times, for the sake of reaching an accurate result and helping you on your way to treatment and recovery.

There is no one, cut and dry, blanket method for treating a goiter, as the treatment will depend entirely on precisely what is the cause that is underlying the goiter. As well, a particular course of action will be decided by your doctor on the basis of the size of the goiter, and the condition that the goiter is in, as well as the symptoms you have that are associated with the goiter. It will also be important to not overlook any factors to your health that may have been responsible for the goiters formation when looking into treatment options.

A goiter which can be regarded as simple, having a prevalence of causing no imbalances in the thyroid gland, as well as no seeming problems as a result of the thyroid gland, will be less likely to cause further obstructions or overall issues.

In order to shrink a goiter, in the case of hypo or hyperthyroidism, it may be enough to just take prescribed medicines as a treatment for the symptoms and for the swelling of the thyroid gland. Medications which are known as corticosteroids often see use in the task of reducing any inflammation, or when there is a prevalence of thyroiditis.

Medicinal treatments for a goiter are not always the most effective response, however. It is not at all uncommon for a goiter to have grown too large to be able to respond properly to medicinal therapy and begin to shrink. In such a case there are surgeries which are available, known as a thyroidectomy. Undergoing a thyroidectomy will mean removing your thyroid gland completely and is a common option for when the thyroid gland grows too large and further obstructs what would otherwise be simple actions, such as swallowing or breathing.

When you are going through the experience of trying to treat what is the most harmful of the goiter family, the toxic nodular or multinodular goiter, RAI, or radioactive iodine treatment is typically the necessary response. You will be given a tablet, the RAI, which is a small amount of the radioactive iodine, which gets ingested orally and begins the process of destroying thyroid gland tissue.

When it comes to the treatment of a goiter, there are options for home care which can be very helpful and ought not to be overlooked as such. When you have finished up with all the treatment that can be offered at the hospital, or by a referred endocrinologist, it is an entirely common possibility that a discussion with your general practitioner will end in him or her suggesting you continue care of yourself in the home, with a prescription of some form of medication, which may end up being a decrease or increase in the amounts of iodine that you are ingesting regularly. This will, of course, be determined by the type of goiter that was ailing you, as well as requiring regular testing to keep an eye on your iodine levels, and the efficiency of your thyroid glands production and secretion of hormones. If it all ends up that a goiter is non-problematic, being too small to count as an issue or cause an imbalance, you may require no treatment or care at home at all.

PART VI

Renal Diet

Chapter 1: Easy Recipes for Managing Kidney Problems

Pumpkin Pancakes

Total Prep & Cooking Time: 40 minutes

Yields: 2 servings

Nutrition Facts: Calories: 183 | Carbs: 39g | Protein: 5.4g | Fat: 1.2g | Sodium: 130mg

Ingredients:

- Two egg whites
- Two tsps. of pumpkin pie spice
- One tsp. of baking powder
- One tbsp. of brown sugar
- Three packets of Stevia
- 1.25 cups of all-purpose flour
- Two cups each of
  - Rice milk
  - Salt-free pumpkin puree

Method:

1. Start by mixing all the dry ingredients together in a bowl – baking powder, Stevia, sugar, flour, and pumpkin pie spice.

2. Now, take another bowl and, in it, mix the rice milk and pumpkin puree thoroughly.

3. In another bowl, form stiff peaks by whipping egg whites.

4. Take the mixture of dry ingredients and add them to the wet ingredients. Blend them in. Once you get a smooth mixture, add the egg whites, and whip them.

5. Grill the mixture on an oiled griddle on medium flame.

6. When you notice bubbles forming on the pancakes, you have to flip them.

7. Cook both sides of the pancakes evenly so that they turn golden brown.

Pasta Salad

Total Prep & Cooking Time: 50 minutes

Yields: 4 servings (half a cup each serving)

Nutrition Facts: Calories: 69 | Carbs: 12.5g | Protein: 2.5g | Fat: 1.3g | Sodium: 72mg

Ingredients:

- A quarter cup of olives (sliced after being pitted)
- One cup of chopped cauliflower
- Two cups of fusilli pasta (cooked)
- Half a unit each of
  - Green bell pepper (sliced)
  - Red onion (chopped)
  - Tomato (small-sized, diced)

Method:

1. Start by cooking the pasta, and for that, you have to follow the directions as mentioned on the package.
2. Now, drain the pasta. Add all the vegetables.
3. Choose any dressing of your choice, but it has to be low-fat. Toss the pasta and the veggies in the dressing.

4. Serve and enjoy!

Broccoli and Apple Salad

Total Prep & Cooking Time: 15 minutes

Yields: 8 servings (3/4 cup each serving)

Nutrition Facts: Calories: 160 | Carbs: 18g | Protein: 4g | Fat: 8g | Sodium: 63mg

Ingredients:

- Four cups of fresh florets of broccoli
- One medium-sized apple
- Half a cup each of
    - Sweetened cranberries (dried)
    - Red onion
- A quarter cup each of
    - Walnuts
    - Fresh parsley
    - Mayonnaise
- Two tbsps. each of
    - Apple cider vinegar
    - Honey
- A three-fourth cup of plain Greek yogurt (low-fat)

Method:

1. Prepare the broccoli florets by cutting into bite-sized chunks. Trim them properly. Take the apple and cut into small pieces as well but in the unpeeled state. Prepare the parsley by chopping them coarsely.

2. Now, take a large-sized bowl and add the mayonnaise, yogurt, vinegar, honey, and parsley. Whisk them together.

3. Take the remaining ingredients and add them too. Make sure they are evenly coated with the yogurt mixture. Once prepared, keep the salad in the refrigerator because it is best served when chilled. It allows the flavors to combine properly. Before serving, stir the salad.

Notes:

- *You can use your favorite type of apple.*
- *If you want, you can sprinkle some more parsley on top just before serving.*

Pineapple Frangelico Sorbet

Total Prep & Cooking Time: 2 hours 10 minutes

Yields: 4 servings

Nutrition Facts: Calories: 119 | Carbs: 28g | Protein: 1g | Fat: 0.2g | Sodium: 2.4mg

Ingredients:

- Two tsps. of Stevia
- One tbsp. of Frangelico (keep two tsps. extra)
- Half a cup of unsweetened pineapple juice
- Two cups of pineapple (fresh)

Method:

1. Take all the ingredients in the container of the blender and process them until you get a smooth mixture.
2. Then, take this mixture and divide it into ice cubes. Keep it in the refrigerator and allow it to freeze.
3. When you find that the mixture has frozen, take them out and blend them in the food processor again. This process will give you a fluffy texture.
4. Before you serve, refreeze the sorbet.

Egg Muffins

Total Prep & Cooking Time: 45 minutes

Yields: 8 servings

Nutrition Facts: Calories: 154 | Carbs: 3g | Protein: 12g | Fat: 10g | Sodium: 155mg

Ingredients:

- Half an lb. of ground pork
- Half a tsp. of herb seasoning blend of your choice
- A quarter tsp. of salt
- Eight eggs (large-sized)
- A quarter tsp. each of
    - Onion powder
    - Garlic powder
    - Poultry seasoning
- One cup each of
    - Onion
    - Bell peppers (A mixture of orange, yellow, and red)

Method:

1. Set the oven temperature to 350 degrees F and use cooking spray to coat a muffin tin of regular size.

2. Prepare the onions and bell peppers by dicing them finely.

3. Take a bowl and in it, combine the following ingredients – garlic powder, poultry seasoning, pork, onion powder, and herb seasoning blend. Form the sausage by combining all of this properly.

4. Now cook the sausage in a non-stick skillet. Once it has been appropriately cooked, drain the sausage.

5. Use salt and milk substitute/milk to beat the eggs in a bowl. In it, add the veggies and the sausage mix.

6. Take the prepared muffin tin and pour the egg mixture into it. You have to leave enough space for the muffins so that they can rise. Bake them for about 20-22 minutes.

Notes: *If there are extra muffins, then you can have them as a quick breakfast the next day, and you simply have to reheat them for about 40 seconds.*

Linguine With Broccoli, Chickpeas, and Ricotta

Total Prep & Cooking Time: 1 hour 5 minutes

Yields: 4 servings

Nutrition Facts: Calories: 404 | Carbs: 49.8g | Protein: 13.2g | Fat: 17.5g | Sodium: 180.4mg

Ingredients:

- Eight ounces of ricotta cheese that have been kept at room temperature
- A bunch of Tuscan kale (chopped into bite-sized chunks and stemmed)
- One-third cup of extra-virgin olive oil
- A pinch of black pepper
- Two cloves of garlic (sliced thinly)
- Fourteen ounces of chickpeas (rinsed after draining)
- Twelve ounces of spaghetti or linguine pasta
- A pinch of kosher salt
- One lemon
- Half a teaspoon of red pepper flakes
- Two tablespoons of unsalted butter
- To taste – Flaky sea salt

Method:

1. Take a large pot and add water to it. Add salt and bring the water to a boil. Cook the pasta by following the directions mentioned on the package. They must be perfectly al dente. Once the pasta is done, you have to drain it but, at the same time, reserve half a cup of the cooking water.

2. Heat the broiler and adjust the rack. Toss the following ingredients together in a bowl – garlic, chickpeas, broccoli, one-third cup of oil, and

red pepper flakes. Everything should become evenly coated. Use pepper and salt to season the mixture.

3. Take a sheet pan and spread the mixture out on it evenly.

4. Take the kale and add it to the previous bowl you used. Toss it again along with the remaining oil, if any. If you need it, then you can drizzle some more oil on top. Spread the kale in a second sheet pan in an even layer.

5. You have to take one sheet at a time while working. Broil the chickpeas and broccoli and halfway through the process, toss them. The broccoli should become charred and tender, and the chickpeas should be toasty. It will take about seven minutes. Then, broil the kale too for about five minutes, and they should become crispy.

6. Take the lemon, zest it, and then cut it into two halves. Take one half and form four wedges out of it. The juice of the lemon will have to be squeezed out on the roasted veggies and then use pepper and salt to season.

7. Place the pasta back in the pot. Take the pasta water you had earlier reserved and add it to the pasta and the lemon zest, butter, and ricotta. Keep tossing so that everything is well incorporated. Now, add the roasted veggies too. If you need, add some more pasta water while tossing.

8. Now, your linguine is done, and you have to divide it among four bowls—season with pepper and flaky sea salt. Squeeze a few drops of lemon on top and serve. If you want, drizzle some more oil before serving.

Ground Beef Soup

Total Prep & Cooking Time: 35 minutes

Yields: 6 servings

Nutrition Facts: Calories: 222 | Carbs: 19g | Protein: 20g | Fat: 8g | Sodium: 170mg

Ingredients:

- Half a cup of onion
- One tbsp. of sour cream
- Three cups of mixed vegetables (frozen, peas, green beans, corn, and carrots)
- One-third cup of uncooked white rice
- Two cups of water
- One cup of beef broth (reduced-sodium variety)
- One tsp. of browning sauce and seasoning of your choice
- Two tsps. of lemon pepper seasoning of your choice
- One lb. of ground beef (lean)

Method:

1. Prepare the onion by chopping them thoroughly. Then, take a large-sized saucepan and, in it, brown the onion and ground beef together. Drain the juices and excess fat.
2. Add the browning sauce and seasonings. Then, add the mixed veggies, rice, water, and beef broth and mix everything together.
3. Bring the mixture to a boil after placing it on high flame. Once the mixture starts boiling, reduce the flame to medium-low and cover the saucepan. Allow it to simmer and cook it for half an hour.
4. Once done, remove the pan from the flame and add the sour cream. Stir it in and serve.

Apple Oatmeal Crisp

Total Prep & Cooking Time: 40 minutes

Yields: 8 servings

Nutrition Facts: Calories: 297 | Carbs: 42g | Protein: 3g | Fat: 13g | Sodium: 95mg

Ingredients:

- A three-quarter cup of brown sugar
- Half a cup of butter
- One tsp. of cinnamon
- Half a cup of all-purpose flour
- Five apples (if possible, then Granny Smith ones)
- One cup of whole oatmeal

Method:

1. Set the temperature of the oven to 350 degrees F and preheat. Peel the apples, core them, and then cut them into slices.

2. Take a bowl and then mix the following ingredients in it together – brown sugar, oatmeal, cinnamon, and flour.

3. Use a pastry cutter to cut the butter into the oatmeal and make sure they are well blended.

4. Take a baking pan of 9 by 9 inches in size and place the sliced apples in it.

5. Take the oatmeal mixture and sprinkle it on top of the apples.

6. Bake the mixture for about thirty to thirty-five minutes.

Chapter 2: Weekend Recipes for Renal Diet

Hawaiian Chicken Salad Sandwich

Total Prep & Cooking Time: 10 minutes + chilling

Yields: 4 servings

Nutrition Facts: Calories: 349 | Carbs: 24g | Protein: 22g | Fat: 17g | Sodium: 398mg

Ingredients:

- One cup of pineapple tidbits
- Two cups of cooked chicken
- One-third cup of carrots
- Half a cup each of
    - Green bell pepper
    - Mayonnaise (low-fat)
- Four units of flatbread
- Half a tsp. of black pepper

Method:

1. Take the cooked chicken and chop it into bite-sized pieces.
2. Prepare the pineapple by draining it and then shred the carrots and chop the bell pepper.
3. Take all the ingredients in a medium-sized bowl and mix them well.
4. Refrigerate the mixture until it is thoroughly chilled.
5. Before serving, spread the chicken on the flatbread's open surface, or if you prefer it wrapped, you can use a tortilla too.

Apple Puffs

Total Prep & Cooking Time: 1 hour 20 minutes

Yields: 12 servings

Nutrition Facts: Calories: 156 | Carbs: 22g | Protein: 1.5g | Fat: 7.3g | Sodium: 176mg

Ingredients:

- Eight ounces of puff dough sheets
- One can (21 oz.) of apple pie filling
- Half a tsp. of rum extract
- One tsp. each of
  - Powdered sugar
  - Baking soda
  - Ground cinnamon

Method:

1. First, you have to thaw the puff dough sheets at room temperature, and it will take you approximately 1 hour.
2. Set the temperature of the oven to 400 degrees F and preheat.

3. Take a bowl, and in it, add the apple pie filling. If you have already sliced the apples, then you can form thirds from them now. Mix the rum extract and cinnamon with the apples.

4. Once the dough has been completely thawed, take one of the sheets and cut nine equal squares from it. Take the other sheet, and you will need only one-third of it to cut another three such squares.

5. Now, take the muffin tin and place the squares in each of the tins. In each of these squares, spoon some of the apple mixture.

6. Bake the preparation in the preheated oven for fifteen minutes, and they should become golden brown in color.

7. Once done, remove the puffs from the muffin tins and before serving, sprinkle some powdered sugar on top of each apple puff. Serve them warm.

Creamy Orzo and Vegetables

Total Prep & Cooking Time: 30 minutes

Yields: 6 servings

Nutrition Facts: Calories: 176 | Carbs: 25g | Protein: 10g | Fat: 4g | Sodium: 193mg

Ingredients:

- Half a cup of frozen green peas
- One tsp. of curry powder
- One carrot (medium-sized)
- One zucchini (small-sized)
- One onion (small-sized)
- One clove of garlic
- Three cups of chicken broth (low-sodium variety)
- Two tbsps. each of
    - Olive oil
    - Fresh parsley
- A quarter tsp. of black pepper
- A quarter cup of Parmesan cheese (freshly grated)
- One cup of cooked orzo pasta
- A quarter tsp. of salt

Method:

1. Start by preparing the veggies. Chop the zucchini and onion. Chop the garlic finely. Then, take the carrots and shred them.

2. Place a large-sized skillet on the oven over medium flame. Heat olive oil in the skillet. Sauté the following ingredients in it for about five minutes – carrots, zucchini, onion, and garlic.

3. After that, add the curry powder to the mixture. Season with salt and then add the chicken broth. Bring the mixture to a boil.

4. Now, add the cooked orzo pasta and keep stirring until the mixture starts boiling. Cover the skillet and allow the mixture to simmer. Keep stirring from time to time and cook the pasta for another 10 minutes. By this time, the pasta will become al dente, and the liquid will be absorbed.

5. Add the chopped parsley, cheese, and the frozen peas into the pasta. Keep heating until the vegetables are sufficiently hot, and if you want to enhance the creaminess, then you can add some more broth—season with pepper.

Minestrone Soup

Total Prep & Cooking Time: 45 minutes

Yields: 4 servings

Nutrition Facts: Calories: 144 | Carbs: 21.9g | Protein: 5.9g | Fat: 4.3g | Sodium: 55.1mg

Ingredients:

- Four cups of low-sodium chicken broth (low-fat)
- One carrot (large-sized)
- One and a half cups of dry macaroni (elbow-shaped)
- 14 oz. of tomatoes (diced, without any salt content)
- Two stalks of celery
- Two garlic cloves
- Half a cup of zucchini (freshly chopped)
- One teaspoon each of
    - Dried basil
    - Dried oregano
    - Freshly ground black pepper
- Half an onion (large-sized)
- One can of green snap beans (without any salt content)
- Two tbsps. of olive oil

Method:

1. Prepare the veggies by dicing zucchini, garlic, and onion. Then, take the carrots and shred them. Either use fresh green beans or canned ones, but you have to cut them into pieces of half an inch size.

2. Take a Dutch oven or a large pot and place it on medium flame—heat olive oil in the pot. Add the diced onions in the pot as well and then cook them for a couple of minutes until they become translucent.

3. Add zucchini, carrot, celery, and garlic, and if you are using fresh green beans, then add them too. Cook the vegetables for about five minutes and they will become tender.

4. Add black pepper, oregano, basil, and if you are using canned beans, then add them now.

5. Add the chicken broth and the diced tomatoes and keep stirring.

6. Bring the mixture to a boil and once it starts boiling, allow the mixture to simmer for about ten minutes.

7. Add the pasta and cook them for an additional ten minutes by following the directions mentioned on the package.

8. Before serving, garnish the pasta with fresh basil on top. Serve into bowls and enjoy!

Frosted Grapes

Total Prep & Cooking Time: 1 hour 5 minutes

Yields: 10 servings (serving size – half a cup)

Nutrition Facts: Calories: 88 | Carbs: 21g | Protein: 1g | Fat: 0g | Sodium: 41mg

Ingredients:

- Three oz. of flavored gelatin
- Five cups of seedless grapes

Method:

1. De-steam the seedless grapes after you have washed them. After that, let them be but make sure they are slightly damp.

2. In a large-sized bowl, add the dry gelatin mix. Remember that you shouldn't be pouring in water.

3. Add these damp grapes into the bowl, and in order to coat them uniformly, toss them well.

4. Now, take a baking sheet, and place these grapes on the sheet in an even layer.

5. Freeze them for 1 hour and then serve chilled.

Notes: *The flavor of the gelatin you use can be adjusted as per your choice. If you want to decrease the carbs, then use gelatin that is sugar-free.*

Yogurt and Fruit Salad

Total Prep & Cooking Time: 2 hours 20 minutes

Yields: 4 servings

Nutrition Facts: Calories: 99 | Carbs: 22g | Protein: 2.6g | Fat: 0.7g | Sodium: 12mg

Ingredients:

- One-third cup of dried cranberries
- Half a cup of pineapple chunks (fresh)
- Six strawberries (large-sized)
- Six ounces of Greek yogurt (strawberry flavored)
- Four ounces of mandarin oranges (drained, light syrup)
- Ten green grapes
- One apple (with skin, medium-sized)

Method:

1. Wash the strawberries, grapes, and apples. After that, pat them dry.
2. Slice the apples and chop them into bite-sized chunks.
3. Then, take the strawberries and slice them as well.
4. Mix the following ingredients together – yogurt, dried cranberries, pineapple, Mandarin oranges, grapes, and apples.
5. Keep the mixture covered and put it in the refrigerator for two hours.
6. Before serving, garnish the preparation with sliced strawberries.

Beet and Apple Juice Blend

Total Prep & Cooking Time: 5 minutes

Yields: 2 servings

Nutrition Facts: Calories: 53 | Carbs: 13g | Protein: 1g | Fat: 0g | Sodium: 66mg

Ingredients:

- A quarter cup of parsley
- Half a beet (medium-sized)
- Half an apple (medium-sized)
- One carrot (fresh, medium-sized)
- One stalk of celery

Method:

1. Process the following ingredients together in a juicer – parsley, celery, carrot, beet, and apple.
2. Take the mixture and pour it into two small glasses. You can either keep the juice in the refrigerator to chill or have it right away.

Notes: *Even though juices are healthy, for kidney patients, you have to be careful so that you don't increase your potassium intake too much.*

Baked Turkey Spring Rolls

Total Prep & Cooking Time: 1 hour 30 minutes

Yields: 8 servings (per serving – 2 spring rolls)

Nutrition Facts: Calories: 197 | Carbs: 9.6g | Protein: 23.3g | Fat: 7.3g | Sodium: 82.2mg

Ingredients:

- 2.5 cups of coleslaw mix
- Two tsps. of freshly ground black pepper
- Twenty ounces of turkey breast (ground)
- Two tbsps. each of
  - Vegetable oil
  - Minced cilantro
- One tbsp. each of
  - Sesame oil
  - Balsamic vinegar
- Two tsps. of freshly ground black pepper
- Sixteen pastry wrappers (frozen spring roll wraps)
- Cooking spray

Method:

1. Set the temperature of the oven to 400 degrees F and preheat.

2. Take the spring roll wrappers out from the freezer so that they can stay under room temperature. Thawing should be done at least half an hour before preparation.

3. Now, take a bowl, and in it, mix the following ingredients with the raw turkey – minced cilantro, sesame oil, and balsamic vinegar.

4. Take a large-sized skillet, and in it, pour two tbsps. of vegetable oil. Put the skillet on medium-high flame and preheat. Add the ground turkey into the skillet and crumble it by stirring. To cook the turkey properly, you have to keep sautéing the mixture.

5. Then, you have to add the mixture of coleslaw to the turkey and keep cooking for another five minutes. Season with freshly ground black pepper – two tsps. should be enough. Mix everything properly.

6. Once done, remove the skillet from the flame. Use a strainer to drain any remaining liquid.

7. Take one spring roll wrapper and near one corner of it – add the filling diagonally. You can take three tbsps. of filling for one roll. There should be adequate space left on both sides. Fold one side towards the inside and do the same with the other side. Roll them and make sure the sights have been tucked in properly. Use water to moisten one of the sides of the wrapper because this helps to seal properly.

8. Take the remaining wrappers and follow the same steps with them.

9. Use non-stick cooking spray to coat the baking pan's base and then place the spring rolls in it. Place the pan in the oven, and it should be complete in half an hour when given at 400 degrees F.

10. You can also serve the rolls with a sweet chili sauce, but this has not been included in the nutrition facts.

Crab-Stuffed Celery Logs

Total Prep & Cooking Time: 10 minutes

Yields: 4 servings

Nutrition Facts: Calories: 34 | Carbs: 2g | Protein: 2g | Fat: 2g | Sodium: 94mg

Ingredients:

- Two tsps. of mayonnaise
- One tbsp. of red onion
- A quarter cup of crab meat
- Four ribs or celery (approx. eight inches in size)
- A quarter tsp. of paprika
- Half a tsp. of lemon juice

Method:

1. Take the celery ribs and trim the ends. Prepare the crab meat by draining it and then use two forks to flake the meat. Chop the onion and mince it thoroughly.

2. Take a small-sized bowl and in it, add the lemon juice, mayonnaise, onion, and crab meat and combine them properly.

3. Take a whole tablespoon full of the mixture and fill the celery rib with it.

4. Each rib of celery has to be cut into three equal pieces.

5. Sprinkle some paprika on top of each of these celery logs.

Couscous Salad

Total Prep & Cooking Time: 50 minutes

Yields: 4 servings (half a cup per serving)

Nutrition Facts: Calories: 151 | Carbs: 28.7g | Protein: 4.9g | Fat: 2.5g | Sodium: 14.3mg

Ingredients:

- One teaspoon each of
  - Dried oregano
  - Allspice
- Two lemons (juiced)
- One tbsp. each of
  - Olive oil
  - Minced garlic
- Half a cup each of
  - Red bell pepper (chopped)
  - Yellow bell pepper (chopped)
  - Carrots (chopped)
  - Frozen corn
- One cup each of
  - Dry couscous
  - Whole sugar snap peas
- Three peeled cucumbers (large-sized)

Method:

1. Follow the package instructions to prepare the couscous. After that, allow it to chill.

2. Take a large bowl and mix the following ingredients: cucumbers, couscous, snow peas, carrots, corn, yellow pepper, and red pepper.

3. Take another bowl of small size and, in it, whisk the following ingredients together – dried oregano, allspice, lemon juice, olive oil, and minced garlic.

4. Combine everything and serve it chilled.

Chapter 3: One-Week Meal Plan

Day 1

Breakfast – Pumpkin Pancakes

Lunch – Ground Beef Soup

Snacks – Frosted Grapes

Dinner – Pasta Salad

Day 2

Breakfast – Yogurt and Fruit Salad

Lunch – Broccoli and Apple Salad

Snacks – Apple Puffs

Dinner – Baked Turkey Spring Rolls

Day 3

Breakfast – Egg Muffins

Lunch – Minestrone Soup

Snacks – Crab-Stuffed Celery Logs

Dinner – Hawaiian Chicken Salad Sandwich

Day 4

Breakfast – Yogurt and Fruit Salad

Lunch – Pasta Salad

Snacks – Apple Puffs

Dinner – Linguine with Broccoli, Chickpeas, and Ricotta

Day 5

Breakfast – Beet and Apple Juice Blend

Lunch – Ground Beef Soup

Snacks – Frosted Grapes

Dinner – Baked Turkey Spring Rolls

Day 6

Breakfast – Pumpkin Pancakes

Lunch – Creamy Orzo and Vegetables

Snacks – Pineapple Frangelico Sorbet

Dinner – Couscous Salad

Day 7

Breakfast – Egg Muffins

Lunch – Broccoli and Apple Salad

Snacks – Pineapple Frangelico Sorbet

Dinner – Ground Beef Soup

Chapter 4: Avoiding Dialysis and Taking the Right Supplements

Even though getting diagnosed with chronic kidney disease (CKD) might appear scary, you can take certain steps to prolong your kidney function and delay the onset of dialysis if you catch the disease in its early stages. Some of the main causes of CKD in Americans are high blood pressure and diabetes. In order to prolong kidney function, these diseases should be controlled.

Steps to Avoid Dialysis and Prolong Kidney Function

There are steps an individual could take to prolong kidney function regardless of how the individual developed CKD.

- **Following a renal diet** – The main aim of a pre-dialysis diet is to maintain optimum nutrition. A renal diet is one that has a low content of protein, phosphorus, and sodium and emphasizes the importance of limiting the intake of fluids and consuming high-quality protein. It's essential to consult your dietician for individualized nutrition counseling. Several doctors believe that the progression of kidney diseases can be slowed down by following a renal diet.

- **Reduce the intake of salt** – Consuming an excess amount of salt with your foods is linked with high blood pressure.

- **Exercise regularly** – Exercises like running, walking, and swimming can help maintain a healthy weight, manage diabetes and high blood pressure, and decrease stress.

- **Reduce stress** – Decreasing stress and anxiety can lower your blood pressure, which in turn can be beneficial for your kidneys.

- **Don't smoke** – Smoking decreases the flow of blood to your kidneys. It decreases kidney function in both people with or without diseases.

- **Limit alcohol intake** – Alcohol consumption can increase your blood pressure. The excess calories can also make you gain weight.

- **Drink enough water** – Your kidneys can be damaged by dehydration, decreasing blood flow to the kidneys. However, follow your nutritionist's guidelines regarding fluid intake because regular fluid intake can also increase the build-up of fluid in your body, which can become dangerous for patients in the later stages of CKD.

- **Control your blood pressure** – High blood pressure can increase your risk of kidney failure and heart diseases.

- **Control your blood sugar** – The risk of kidney failure and heart diseases are increased due to diabetes.

- **Maintain a healthy weight** – The risk of kidney-related conditions like high blood pressure and diabetes can be increased because of obesity.

Even though CKD cannot be reversed, appropriate treatment can slow down its progression. See your doctor regularly to monitor your kidney function and slow the progression of kidney failure.

Supplements to Look Out for

The dietary requirements of people who are suffering from any sort of kidney problems are not always the same. Someone might need extra calories and proteins, whereas others might need fewer amounts of such nutrients. Thus, a professional healthcare provider is the best person who can assist and guide you for choosing the perfect supplements necessary for your kidney disease. Special supplements meant for keeping the kidney safe are available in various sizes, shapes, flavors, and forms. It is always necessary to consult a healthcare practitioner before consuming any nutritional supplement related to the kidney.

Individuals who are suffering from chronic kidney disease (CKD) require certain water-soluble vitamins in higher quantities. Here you will get to know about some of the supplements that are meant for dealing with kidney problems.

- **Vitamin B1 or Thiamin -** It looks after the proper functioning of the nervous system. Thiamin also helps the cells in producing the required amount of energy from carbohydrates. People with chronic kidney disease are recommended to intake 1.5mg of this water-soluble vitamin supplement per day.

- **Vitamin B2 or Riboflavin -** Vitamin B2 supports healthy skin as well as normal vision. People who are fighting against CKD and are also following a special low-protein diet might consume 1.8mg of Riboflavin supplement each day. Those of you who have a low appetite and are pursuing dialysis might take 1.1 to 1.3mg of vitamin B2 supplements per day.

- **Vitamin B6 -** This effective water-soluble vitamin helps produce proteins that are further used for making cells. Patients of CKD who are under dialysis treatment might consume 10mg of this supplement each day. Those who are non-dialysis patients are recommended to intake 5mg vitamin B6 supplements every day.

- **NAC -** NAC or N-acetylcysteine is an essential amino acid that generally targets the oxygen radicals. Various findings and researches suggest that NAC supplementation is beneficial for hemodialysis patients. NAC supplement decreases oxidative stress as well as improves results of uremic anemia, which is a problem of CKD.

- **ALA -** The antioxidant Alpha lipoic acid might prove helpful in treating certain complications of kidney disease. Supplementation of ALA enhances the action of a few antioxidant enzymes. Such enzymes protect against oxidative disorders and stress.

- **Vitamin B12 -** Vitamin B12 maintains the nerve cells and, in association with folate, produces red blood cells. Both dialysis and non-dialysis CKD patients are recommended to intake 2-3 mg of this supplement per day. Its deficiency can result in permanent nerve damage.

Supplements for kidney problems are better to consume only if it is approved or prescribed by your doctor.